Lecture Notes in Social Networks

Lecture Notes in Social Networks (LNSN) comprises volumes covering the theory, foundations and applications of the new emerging multidisciplinary field of social networks analysis and mining. LNSN publishes peer- reviewed works (including monographs, edited works) in the analytical, technical as well as the organizational side of social computing, social networks, network sciences, graph theory, sociology, semantic web, web applications and analytics, information networks, theoretical physics, modeling, security, crisis and risk management, and other related disciplines. The volumes are guest-edited by experts in a specific domain. This series is indexed by DBLP. Springer and the Series Editors welcome book ideas from authors. Potential authors who wish to submit a book proposal should contact Annelies Kersbergen, Publishing Editor, Springer

e-mail: annelies.kersbergen@springer.com

Song Yang
Editor

Social Network Analysis in Action

Basic Methods and Applications

Springer

Editor
Song Yang
Department of Sociology and Criminology
University of Arkansas
Fayetteville, AR, USA

ISSN 2190-5428 ISSN 2190-5436 (electronic)
Lecture Notes in Social Networks
ISBN 978-3-031-66660-5 ISBN 978-3-031-66661-2 (eBook)
https://doi.org/10.1007/978-3-031-66661-2

This work was supported by National Science Foundation.

The authors of chapters 5 and 6 appreciate funding supports from National Science Foundation under Award No. OIA-1946391 that facilitates development of these chapters. Any opinions, findings, and conclusions or recommendations expressed in this material are those of the authors and do not necessarily reflect the views of the National Science Foundation.

This Springer imprint is published by the registered company Springer Nature Switzerland AG
The registered company address is: Gewerbestrasse 11, 6330 Cham, Switzerland

Contents

About the Editor

Song Yang is professor of Sociology at the Department of Sociology and Criminology, University of Arkansas. His research emphasizes social network analysis (SNA), focusing on SNA methodology and its application to analyzing wide range of research topics. He published more than 20 peer reviewed articles and several books, such as *Social Network Analysis* (with David Knoke, 2008) and *Social Network Analysis* (with David Knoke, 2020), and *Social Network Analysis: Methods and Examples* (with Franziska Keller and Lu Zheng, 2016).

Chapter 1
Social Network Analysis: An Introduction

Song Yang

Abstract This chapter starts with a brief historical review of Social Network Analysis (SNA), tracing its root to Sociology studies, while emphasizing its strong theoretical motivations. It includes a description of methods, clarifying basic differences between ego-centric and full network analysis. It discusses SNA history and their applications on different subjects. It concludes with a section to discuss the organization of this book, clarifying chapters and their topic coverages.

Keywords Social network analysis fundamentals · Social network analysis theories · Social network analysis histories

This chapter starts with a brief historical review of Social Network Analysis (SNA), tracing its root to Sociology studies, while emphasizing its strong theoretical motivations. It includes a description of methods, clarifying basic differences between ego-centric and full network analysis. It discusses SNA history and their applications on different subjects. It concludes with a section to discuss the organization of this book, clarifying chapters and their topic coverages.

A social network is a social structure made up of a set of social actors (such as individuals, organizations, neighborhoods, parties, or nation/states) and a set of ties, relations, connections, or interactions between those actors. Social network analysis examines social relations, structures, and networks between various social entities such as individuals, teams, groups, parties, organizations, and nation/states [20]. Social network analysis is often defined with technical terms, such as a set of quantitative or qualitative methods being applied to analyzing relations between a set of finite number of cases. However, social network analysis is much more than a set of methodological tools. It builds around three pillars: theory, methods, and applications.

S. Yang (✉)
Department of Sociology and Criminology, University of Arkansas, Fayetteville, AR, USA
e-mail: yangwang@uark.edu

S. Yang (ed.), *Social Network Analysis in Action*, Lecture Notes in Social Networks,
https://doi.org/10.1007/978-3-031-66661-2_1

1.1 Social Network Analysis Theories

In theory, social network scholars have been very serious about theorical aspects of social networks, focusing on the synergy of the parts and system outcomes at aggregation level. In particular, a simple summation of parts does not produce its arithmetic sum-up at the aggregate level. The systemic outcome of the summation of a social network is either larger or smaller than its arithmetic summation, depending on the relationship between parts. For example, a fruitful collaboration between a set of R & D scientists produce innovative products that none of those scientists could have developed on they own. Conversely, separations or even isolations between members of a group of genius innovators result in low number of innovations being produced. Figure 1.1 illustrates the differences between the two social network configurations and their respective outcomes.

Social network analysis is central ingredient to Mark Granovetter's [11] classical piece on social embeddedness and economic actions, which laid foundation for the neoclassical socioeconomics. Granovetter [11] contended that social network analysis can be used to explain and analyze economic behaviors and actions. In particular, he criticized economics under socialized view of human economic behaviors. He also surprisingly disapproved sociologist over socialized view of human behaviors. He cited Duesenberry's [7] quip that "economics is all about how people make choices, sociology is all about how they don't have any choices to make" to debate that neither approach is ideal to analyze social economic actions ([11]: 485). Social network analysis affords key to avoid the over-socialized view or under socialized view, striking a balance between the two extreme approaches. In particular, social network analysis emphasizes one's individual social contexts in explaining his/her behaviors, avoiding the over socialized view of a sweeping argument that people merely respond to their cultures and norms. Similarly, examining one's social networks also enables researchers to investigate the individual's micro-social environment, avoiding the under socialized view premising on a set of fixed assumptions of human behaviors invariantly applied to all segments of population.

So other than the utility of avoiding over socialized and under socialized view in analyzing social economic behaviors, does social network analysis have any theoretical assumptions on its own? To this, Knoke and Yang [12] discussed three theoretical underpinnings propagating in social network analysis. First, theoretical relations are often more important for understanding observed behaviors than are individual attributes such as, gender, race, age, education occupation, income, and personal temperaments. For example, one's political view is very plastic and difficult to gauge. Often the political views are sets of hybrid forms of beliefs, supports, and agreements that are issue-specific, rather than being strictly partisan that is either all Republican or all Democrats. In addition, the increasing number of undecided voters in key battle ground during crucial elections add tremendous challenges to the task of political analysis. Political scientists have hard time explaining the political views with traditional atomistic variables looking at only the individual attributes (race, gender, social economic status etc). Such challenge compels political scientists to

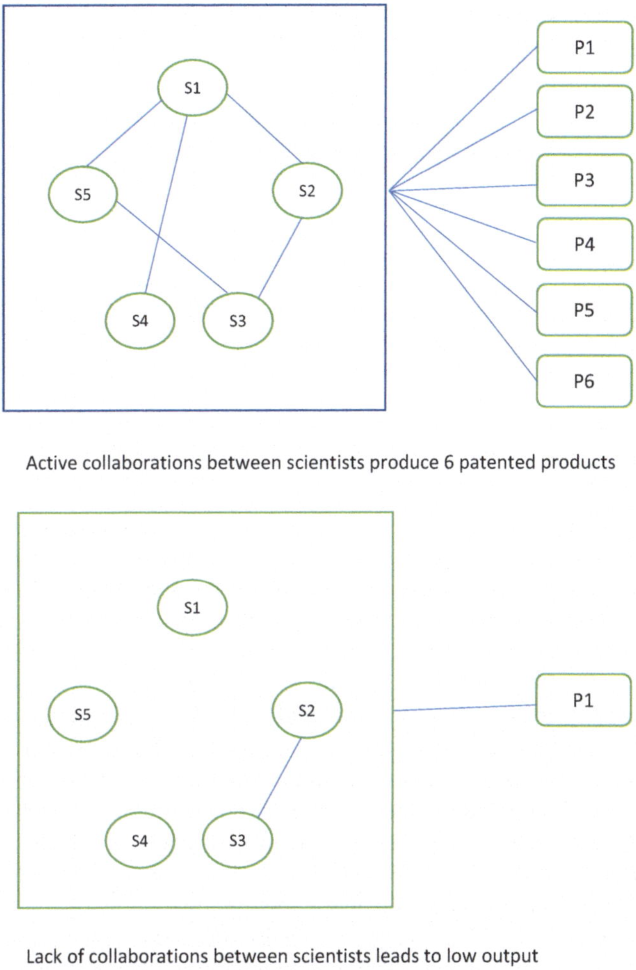

Fig. 1.1 Two social networks and their systemic outcomes

look beyond the individual characteristics, taking serious considerations of one's social network. Social network analysis of political actions yields fruitful results. For example, Nickerson [18] reported that each act of voting of an individual on average generates three additional votes from his/her social contacts. Bond et al. [3] conducted a large-scale voter turnout experiment with Facebook platform, reporting that those who received information about their friends' voting via Facebook postings are more likely to vote and to seek political information than those who received none of such information.

Second, social networks affect an individual's perceptions, beliefs, and actions through diverse structural mechanism. For example, the famous thesis of structural holes by Ron Burt [4] asserted that individuals (as well as corporate actors) can

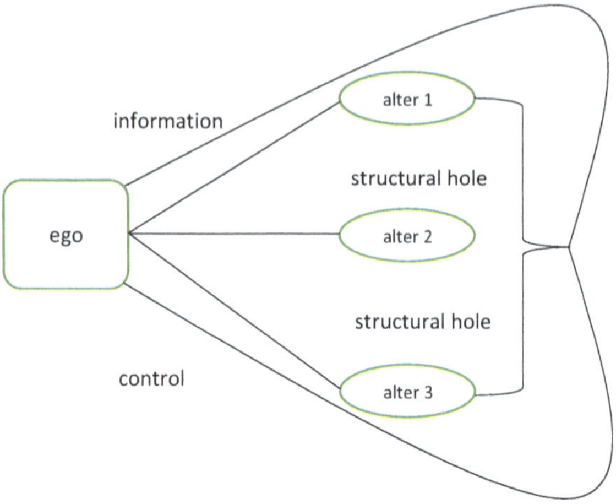

Fig. 1.2 Structural hole and its benefits

benefit from having structural holes in their network configurations via connections to other individuals that are disconnected among themselves. In other words, the disconnected actors provide structural holes, and individuals reap benefits by crossing over such structural holes via connections to those disconnected actors. So, what are those benefits that come with crossing over the structural holes? Per Ron Burt, they are control and information benefits. Compared with individuals who have social closures in their networks, individuals who have structural holes receive novel, additive, and nonredundant information from their diverse sources of network contacts. Individuals who crossover structural holes can also exert control benefits by playing their disconnected contacts off each other, deciding on when and how much information they can brokerage between their contacts. Figure 1.2 displays how the information and control benefits materialize to individuals who crossover structural holes among its social contacts.

However, empirical evidence suggests that the alleged benefits of structural holes only accrue to men, not women [5]. For women, social closures, characterized by strong and redundant ties connecting a cluster of friends afford benefits in the forms of large pay raise and early career promotion. Burt [5] proposed two explanations for such gender disparity: (1) cultural explanation that women tend to thrive in a dense and supportive social environment. While men are very adaptable with leveraging structure holes among their contacts, women may not be comfortable with playing their contacts off each other to exact benefits. (2) institutional explanation that women facing glass ceiling need to bundle together to form a collective force strong enough to break the institutional barrier. Figure 1.3 shows the process whereby social closure, not structural hole affords benefits to women.

Third, the structural mechanism that sustains a network is dynamic, changing, and evolving out of its components' purposive or unpurposive actions. Marriage between two nice couples reflect progressive movement of relationship over time. Unfortunately, a significant portion of marriages ends in divorces between the same nice couples due to relationship deteriorates or other distractions. Inter-

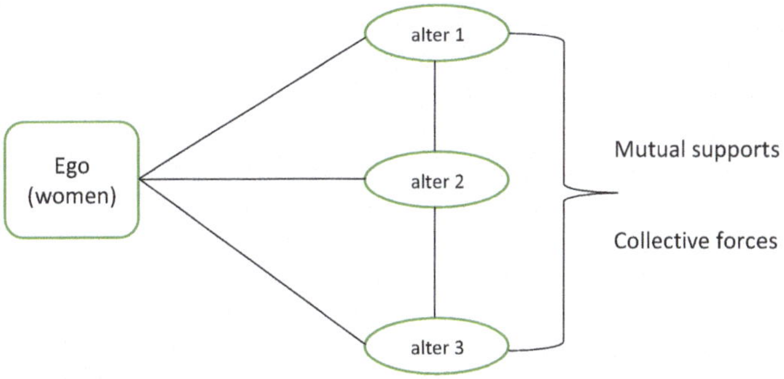

Fig. 1.3 Social closure and its benefits

organizational relationship often starts with the contracts ridden with elaborated legal terms and protections, but with the progress of relationship, such contracts give away to more relaxed informal interactive management. Much like marriages between people, inter-organizational relations also can dissolve due to disagreements, disputes, or other issues. One such example is the joint venture between Motorola and Panda Electronic Group of Nanjing China in 1995. Motorola seeks to expand its market to China, whereas Panda Group needs technology and capital investment from Motorola. However, this joint venture is short lived, dissolving in 1997 when Motorola failed to reach a license agreement with Apple Computer [13, 21].

1.2 Social Network Analysis: A Typology

Social network analysis has a great deal of methods to offer analyzing various types of social networks. However, before I delve into the details of those analytical methods, I need to clarify a fundamental difference between two types of social networks analysis, based on mostly how the data are collected. Such difference is very consequential to the discussions of their respective analytical methods; thus, such clarification prevents any confusions down the road.

Ego-centric social network analysis, sometimes called egonet is based on social network data collected with conventional social science surveys. The famous General Social Survey (GSS) 1985 and GSS 2004 illustrated such approach. In particular, GSS is a nationally representative sample of US adult population to gauge the public opinions and responses to basic demographic features, and a wide range of social issues annually or bi-annually since 1972. In 1985 and 2004, GSS added module questions to capture people's social network with the following measuring question:

> From time to time, most people discuss important matters with other people. Looking back over the last six months-who are the people with whom you discussed matters important to you? Just tell me their first names or initials

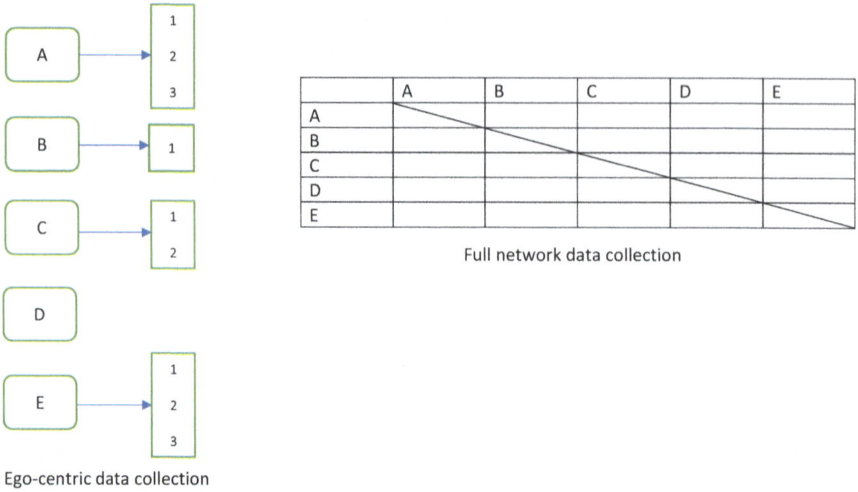

Fig. 1.4 Ego-centric and full network data collection

Such simple design generates very powerful results that reveal a great deal of American's social networking in terms of with whom they discuss important matters. For example, in 1985, average Americans have 3 conversational partners to discuss important matters. Those conversational partners tend to be at the same level of education, to be in the same racial and ethnicity group, and very balanced involving both genders in the network. Main reason behind this finding is significant kinship representation in American's ego-centric network [15]. Nearly 20 years later, American's core discussion group size shrank to 2, with gender, race, education, and occupational compositions exhibiting very little changes from 1985 to 2004 [16]. Such finding leads to claim of social isolation of Americans, triggering a great deal of debates about the veracity of the finding. While some scholars embrace the finding as supportive of their long-term observation of decline of social and civic engagement of Americans, others noted about the online activities, interview effects, or even training of interviewers that cast doubts on the finding.

Key feature in ego-centric network design, in comparison to the full network studies, is that researchers do not draw boundaries to collect data, ego-centric network data are collected along with other social variables in an "open population survey." Consider this artificial example, one researcher studies student's friendship network in a classroom of five students (A) by asking students to name three of their best friends or (B) by asking students to name their three best friends within the same classroom of 5 students. The A approach follows the ego-centric network study approach, and the B approach follows a full network approach. Figure 1.4 illustrates such difference.

Graph on the left of Fig. 1.4 shows the ego-centric design to collect data "name three of your best friends." Each ego (A, B, C, D, and E) nominates friends: A has three, B has one, C has two, D has none, and E names three. Those nominations occur without any boundary, respondents may think of their childhood friends, workplace buddies, church goers, or bar hoppers. The persons being nominated by different egos have little chance to know each other, although those named by the same ego may very well know each other due to their connections with the same ego.

In contrast, full network design imposes boundary to the nominations, asking respondents to restrict their nominations within a preset boundary. Figure 1.4 illustrates the same questionnaire "name three of your best friends." However, the key difference here is that the nominations must be made within the same classroom. For each respondent, the persons they named may very well be very different from the persons they named in the ego-centric design. Such difference is not due to the true aspects of their personal network, but due to the research design. For researchers, understanding such difference is crucial because we must understand the social context in which the data are collected to interpret the analytical results accurately.

The design difference between ego-centric and full network analysis is also very consequential to the analytical methods following the initial data collection. Ego-centric network data are stored and processed in the conventional social science data format with rows representing individual cases and columns representing variables of the cases. Those individual cases are from the open population surveys with very little chance to know each other. The variables of ego-centric network include respondents variables, their contacts' variables, and variables that capture the relationship features between egos (respondents) and their alters (contacts). Analyzing ego-centric network is pretty much the same as analyzing other aspects of social surveys. Marsden [15] and McPherson et al. [16] classical pieces on American's core discussion network illustrates the ego-centric network from design to implementation and to data analysis.

In contrast, the data format for full network analysis are matrices, either in the form of squared matrices or bipartite matrices. The squared matrices are presented with the set of nodes in rows (Xs) and columns (Ys), and entries of the joints between X_i and Y_i indicating the relationship between X_i and Y_i. Using the previous example of "naming your three best friends in the class," I conjure up a hypothetical matrix and its corresponding figure shown in Fig. 1.5.

Looking at the matrix on the left, Actor A has two friends (Actors B and D). Actor B has two friends (Actors A and E). Actor C has one friend (Actor D). Actor D has one friend (Actor C), and Actor E has one friend (Actor B). However, reading across rows and columns for a given network can be very tiresome and risk of seeing trees without seeing the forest. To facilitate communications with readers, network scholars often use figure to illustrate the personal network. Looking at the figure on the right, the lines indicate the friendship (the absence of line suggests no friendship). Corresponding to the matrix on the left, the figure on the right shows A is a friend with B and D, B is a friend with A and E. C is a friend with D, D is a friend with A and C, and finally E is a friend with B.

The social network depicted in Fig. 1.5 is the simplest one compared with others with more variations in values to the relations and directions of the relations. Noted that in Fig. 1.5, all the relations are mutual (un-directed) and binary (presence/absence of the ties). Those assumptions need to be relaxed to accommodate many

	A	B	C	D	E
A	0	1	0	1	0
B	1	0	0	0	1
C	0	0	0	1	0
D	1	0	1	0	0
E	0	1	0	0	0

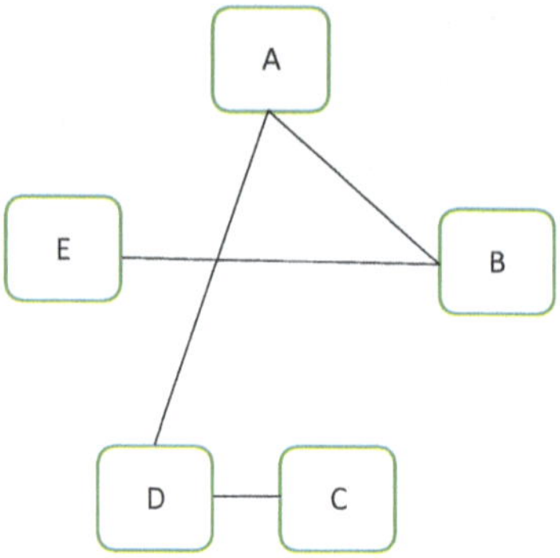

Fig. 1.5 A hypothetical example of full network matrix and figure

Table 1.1 A typology of social networks

Direction Valuation	Undirected	Directed
Binary	I: Undirected binary	II: Directed binary
Valued	III: Undirected valued	IV: Directed valued

realistic occurrences. For example, in communication network, messages are sent from one actor to another in different amounts and message exchanges may not be mutual between pairs of actors. Accommodating both changes in relational direction and intensity, I come up with a typology that contains for types of social networks. Table 1.1 shows such typology.

The social network in Fig. 1.5 corresponds to Type I in Table 1.1 as the relations is all mutual, and valuations are binary (presence/absence of ties). Type II in Table 1.1 refer to social networks with relationship that is directed and binary. Think about a social network in which individuals nominate others for a certain position. Nominations are anonymous, and clear difference exists between those who nominator and those who receive nominations. Type III in Table 1.1 is about social networks that are valued and undirected. A friendship network in realistic can be in such type as friends are distinguished between acquaintance, close friends, and confidants. Type IV in the table is the most complex, but real-life examples abound. As previously mentioned, communications networks are often directed and valued, international trade networks are also directed and valued. In today's globalized world, it is very difficult to find any nations that are completely outside of the international trade relations and networks. Hence, a binary picture of the global trade network is insufficient to depict the trade relations between all nations. International trade relation is also very directed, according to US Census Bureau [19] (https://www.census.gov/foreign-trade/balance/c5700.html), In 2021, US exports $151,442.2 millions of goods to China, while importing $504,935.4 millions from

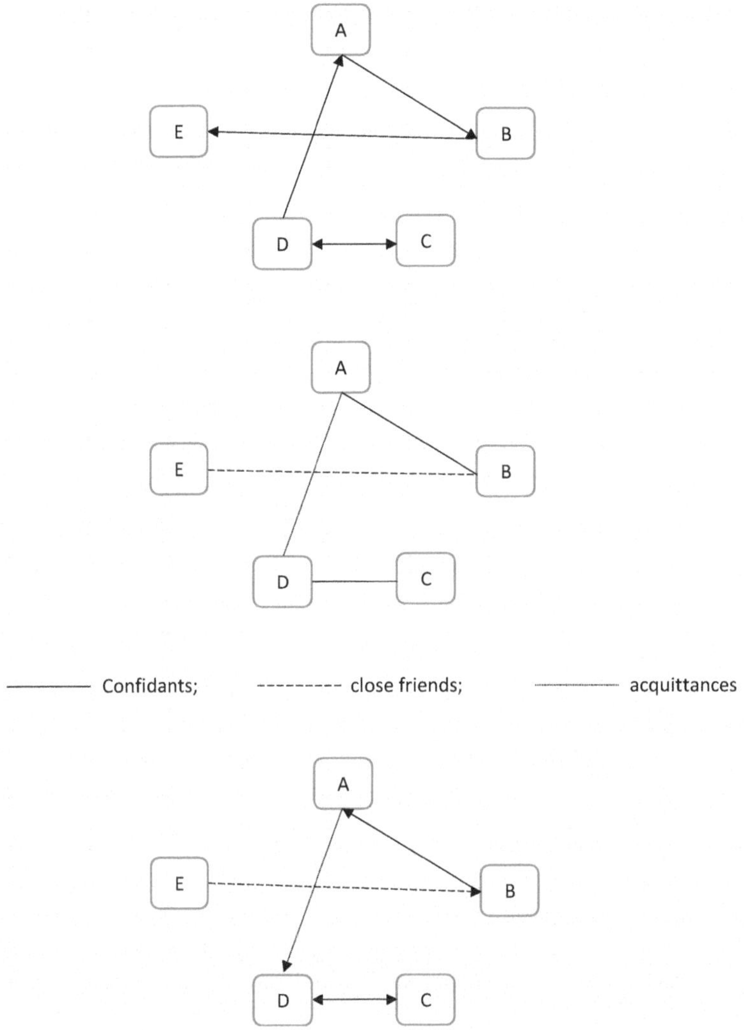

Fig. 1.6 Realistic occurrences of social network typology

China, incurring a trade deficit of $353,493.2 millions. Clearly, the trade relation between the world two largest economies is not symmetric, with US being the buyer of Chinese goods and China being the seller to US market. Fig. 1.6 illustrates the three types of social networks, with Type II (binary directed) above, Type III (valued undirected) in the middle, and Type IV at the bottom (valued directed).

1.3 Social Network Analysis: History and Applications

Social network analysis predated modern day social media such as Facebook and Twitter that takes social network to the online platforms (for details of development of social network analysis in sociology, see [9]). As early as nineteenth century, Sociologists such as Georg Simmel, Emily Durkheim, and Max Weber adopted structural analysis of human behaviors, laying solid foundation for social network analysis in sociology. Later, social psychologist Jacob Moreno examined individual's psychological well-being, and related it to their social connections with others. Moreno and Helen Jennings developed a technique of sociometry to provide a visual illustration that reveals the hidden structure of complex interpersonal networks.

Unfortunately, right after Moreno and Jennings' breakthrough work with sociometry, social network analysis went into a dark age, during which social network perspectives were not the focus of social sciences [9]. Instead, functionalism was prevalent in sociology, which emphasizes shared norms and beliefs and social functions each part of a society plays. Functionalism, being another branch of social structural theory, makes a sweeping assumption of social behaviors being a byproduct of one's social norms and beliefs, which contributes to the overall function of a society. By doing so, it neglects the microlevel social environments conditioned by individuals' social network that produce their social behaviors [11]. An exceptional figure during this dark time of social network analysis was Edward Laumann, a Harvard trained social network scholar working at the time at the University of Michigan. Laumann continued social network analysis of politics, sexual behaviors, and stratification. He trained many doctoral students such as Ronald Burt, Peter Marsden, Joseph Galaskiewicz, and David Knoke, who adopted social network analysis as the central paradigm in their research agenda.

Social network analysis is revived in the 1970s with Harrison White's breakthrough work on structural equivalence and blockmodeling, two important social network analysis techniques that are still widely used today. White produced many doctoral students who are now distinguished scholars in the area of social network analysis, Peter Bearman, Ronald Breiger, Philip Bonacich, Kathleen Carley, Bonnie Erikson, Claude Fischer, Mark Granovetter, and Barry Wellman. Meanwhile, at the University of California Irvin (UCI), James March was building a new institute, which leverages expertise from social sciences and STEM disciplines. When Linton Freeman took over the institute, he quickly built a doctoral degree concentrating in social network analysis, while drawing expertise from sociology as well as mathematical science. The UCI institute soon became the hub of social network analysis, attracting renowned scholars as well as educating a new cohort of social network scholars with valuable interdisciplinary skillsets of social sciences and mathematics, Martin Everett, Tom A.B. Snijders, Stanley Wasserman, Stephen Borgatti, and Philippa Pattison.

Social network analysis continues to benefit from multidisciplinary efforts to push its methodology. For years, social network analysis has been stellar in decoding complex social world and representing it with straightforward visualized social graphs [17]. However, social scientists, with their tradition to uncover the root causes of social phenomenon, are not content with only visualizing what has happened. They want to understand why things happened in the way they did, and the new method of exponential random graph modeling (ERGM) arrived to address the why questions. ERGM is one of the major breakthroughs in recent social network methodologies that benefits from collective efforts from social scientists, mathematicians, statisticians, and computer scientists [14]. ERGM is able to explain the emergence of network ties with individual actor attributes, while controlling for the effects from the endogenous network processes (exp. reciprocity and transitivity). By doing so, it accounts for the issues of actor interdependence in network data. At the initial stage of ERGM development, researchers have been developing and using different packages. Currently, it appears that the Statnet package in R affords the industrial standard for ERGM analyses [19].

Much like other social science research, social network analysis encounters some challenges during the historically unprecedented Covid-19 pandemic. Those challenges mostly lie in the data collection that many social network data, especially the small interpersonal network, requires either in person observation or in person interviews. Such mode (in personal) of data collection was unattainable during the Covid, and the response rate continues to decline even after the worst of Covid-19 rampant [6].

However, such challenges also present opportunities that online social media network sites flourished. US adults using at least one social media (Facebook, YouTube, Twitter, Snapshot, WhatsApp, TikTok etc) to network with others increase from 5 percent in 2005, to 25 percent in 2008, to 50 percent in 2011, and to 75 percent in 2020 [8]. For long, online social media network present a treasure trove to social network scholars to mine the massive online datasets. Online social networking presents a candid shot of individual's network activities, without relying on person's self-report of social networking, which is susceptible to forget and false recall [10]. The challenges lie in the lack of technology background or training for social scientists to mine and scrape these massive online data. Fortunately, recent years witness increasing collaborations of social scientists and data scientists, as well as commercialized packet that can be tweaked to address various research topics. One such effort is the use of Application Programming Interface (API) in text mining of social media data [1, 2].

1.4 Organizations of the Book

This book has six chapters, starting with this Chap. 1. Chapter 2 covers basic research design components of SNA, starting with sampling, boundary specifications, data collection procedures. It also describes online data mining, Application Programming Interface (API), and web scrapers. It introduces a particular engineer product called Intelligent Cyber Argumentation System (ICAS), which was designed and implemented by an interdisciplinary team of computer engineers and social scientists at University of Arkansas through their NSF funded project Data Analytics that are Robust and Trusted (DART).

Chapter 3 covers the basic topic of network representation with graphs and matrices. It introduces elementary methods for analyzing nodes, such as centrality, power and prestige. It describes distance and reachability measures for dyadic pairs, cliques and components for subgroup partitions, size, density, and centralization for analyzing the whole network, structural, regular and automorphic equivalence for analyzing dads. Chapter 4 is the continuation of Chapter 3 to cover advanced social network analytical methods, such as Multidimensional scaling (MDS), bipartite network, and Exponential Random Graph Modeling (ERGM).

Chapters 5 and 6 in this book illustrate use of a data mining product called "the Intelligent Cyber Argumentation System (ICAS)" to facilitate data collection and data analytics of social networking among hundreds of students enrolling in an Introduction to Sociology class. The development of ICAS and applying it to a classroom for social network data collection reflect concerted efforts from data scientists and social scientists, thanks to a large NSF grant that sponsor a project called Data Analytics that are Robust and Trusted (DART) (https://dartproject.org/). In this collective efforts, social scientists design questionnaires and controversial social topics; data scientists develop the ICAS system, which was deployed to the students in a large section of Introduction to Sociology class with the social issues questions loaded as prompt to students. Details of the system, and its data structure and analytics are forthcoming in Chaps. 5 and 6.

References

1. Adams J (2020) Gathering social network data. Sage, Thousand Oaks
2. Aggarwal C, Wang H (2011) Text mining in social networks. In: Aggarwal C (ed) Social network data analytics. Springer, Boston. https://doi.org/10.1007/978-1-4419-8462-3_13
3. Bond RM, Fariss CJ, Jones JJ, Kramer ADI, Marlow C, Settle JE, Fowler JH (2012) A 61-million-person experiment in social influence and political mobilization. Nature 489(7415):295–298. https://doi.org/10.1038/nature11421
4. Burt RS (1992) Structural holes: the social structure of competition. Harvard University Press. ISBN 978-0-674-84371-4
5. Burt RS (1997) The contingent value of social capital. Adm Sci Q 42(2):339–365. https://doi.org/10.2307/2393923
6. Davern M, Bautista R, Freese J, Morgan S, Smith TW (2021) General social survey
7. Duesenberry J (1960) Comment on "An economic analysis of fertility." In Demographic and economic change in developed countries, edited by the Universities-National Bureau Committee for Economic Research. Princeton: Princeton University Press
8. Entwisle B (2020) Series editor introduction, pp 1–2 in David Knoke and Song Yang, Social network analysis, 3rd edn. Quantitative applications in the social sciences, No. 154. Sage Publications, Thousand Oaks
9. Freeman LC (2004) The development of social network analysis: a study in the sociology of science. Empirical Press, Vancouver
10. Freeman L, Romney K, Freeman S (1987) Cognitive structure and informant accuracy. Am Anthropol 89(2):310–325. https://doi.org/10.1525/aa.1987.89.2.02a00020

11. Granovetter M (1985) Economic action and social structure: the problem of embeddedness. Am J Sociol 91:481–493
12. Knoke D, Yang S (2020) Social network analysis. Sage, Thousand Oaks
13. Luo Y (2000) Entering China today: what choices do we have? J Glob Mark 14(1):57–82
14. Lusher D, Koskinen J, Robins G (2012) Exponential random graph models for social networks: theory, methods, and applications. Cambridge University Press, Cambridge
15. Marsden PV (1987) Core discussion networks of Americans. Am Sociol Rev 52:122–131
16. McPherson M, Brashears M, Smith-Lovin L (2006) Social isolation in America: changes in core discussion networks over two decades. Am Sociol Rev 71(3):353–375. https://doi.org/10.1177/000312240607100301
17. Morimoto SA, Yang S (2013) What friendship entails: an empirical analysis of graduate students' social networks. Sociol Spectr 33(2):99–116. https://doi.org/10.1080/02732173.2013.732870
18. Nickerson DW (2008) Is voting contagious? Evidence from two field experiments. Am Polit Sci Rev 102(1):49–57. https://doi.org/10.1017/S0003055408080039
19. Vega Yon G G, Slaughter A, de la Haye K (2021) Exponential random graph models for little networks. Soc Netw 64:225–238
20. Yang S (2022) Social network analysis. In: Frey BB (ed) The SAGE encyclopedia of research design. Sage, Thousand Oaks, pp 1546–1551
21. Yang S, Keller FB, Zheng L (2016) Social network analysis: methods and examples. Sage, Thousand Oaks
22. US Census Bureau. https://www.census.gov/foreign-trade/balance/c5700.html. Downloaded April 1 2024

Chapter 2
Social Network Analysis: Data Collections

Song Yang

Abstract Before social network analysts can start to examine social networks, one critical task is to collect social network data. Where do social network data come from? What steps we need to take to collect social network data? This chapter strives to answer those questions, it begins with a clarification note of the key differences between two distinctive social network analysis design: ego centric network (or ego-net) and whole network studies. It then moves on to describe topics of sampling and boundary specification in collecting the whole social network data. It also discusses methods to mine online social media data with Application Programming Interface (API), and web scrapers. This chapter ends with an introduction of a particular engineer product called Intelligent Cyber Argumentation System (ICAS), which was designed and implemented by an interdisciplinary team of computer/data scientists/engineers and social scientists through an NSF funded project Data Analytics that are Robust and Trusted (DART: https://dartproject.org/).

Keywords Social network data collections · Online data mining · Application programming interface and web scrapping · Intelligent Cyber Argumentation System (ICAS)

Before social network analysts can start to examine social networks, one critical task is to collect social network data. Where do social network data come from? What steps we need to take to collect social network data? This chapter strives to answer those questions, it begins with a clarification note of the key differences between two distinctive social network analysis design: ego centric network (or ego-net) and whole network studies. It then moves on to describe topics of sampling and boundary specification in collecting the whole social network data. It also discusses methods to mine online social media data with Application Programming Interface (API), and web scrapers. This chapter ends with an introduction of a particular engineer product called Intelligent Cyber Argumentation System (ICAS), which was

S. Yang (✉)
Department of Sociology and Criminology, University of Arkansas, Fayetteville, AR, USA
e-mail: yangwang@uark.edu

S. Yang (ed.), *Social Network Analysis in Action*, Lecture Notes in Social Networks,
https://doi.org/10.1007/978-3-031-66661-2_2

designed and implemented by an interdisciplinary team of computer/data scientists/ engineers and social scientists through an NSF funded project Data Analytics that are Robust and Trusted (DART [5]: https://dartproject.org/).

2.1 Ego Centric Network and Whole Network: Two Distinctive Approaches

Social network analysis benefits greatly from collective efforts from a diverse discipline involving eclectic methods, design, and general approaches. While such diversity affords social network analysis with great strength, it can also be a source of confusion, producing many challenges to social network scholars who must scramble and negotiate through such diverse area of study. To reduce the level of perplexity in social network analysis, this chapter provides a critical clarification of the two distinctive social network analysis approaches between ego-net and whole network analysis.

Ego-net is a set of relationship, connection, or ties revolving around an actor, who is also called ego or focal individual. Ego-net analysis is a set of design and methods developed to collect and analyze the ego-net: the social networks around the ego/focal individuals. The sampling of ego-net data is essentially the same process as the sampling of other conventional social surveys. So, the key in ego-net data collection is not about sampling design (as it follows the conventional sampling), but rather it lies in the survey instruments used to capture the ego-net. The instruments used in those surveys vary, but three common types emerged in the history of catching ego-net: name generators [24, 26], positional generators [18], and resource generators [34]. In ego-net surveys, those individuals in the survey being asked of their personal networks are called egos, and their named social contacts are called alters. Sections below describe in detail of the three generators in ego-net data collection processes.

2.2 Name Generator, Positional Generator, and Resource Generator

2.2.1 Name Generator

Name generator has an established history in social research as it was implemented in two waves of General Social Surveys (GSS). Starting in GSS 1985, respondents in the survey were asked the following questions [24]:

> From time to time, most people discuss important matters with other people. Looking back over the last 6 months—who are the people with whom you discussed matters important to you? Just tell me their first names or initials.

The interviewer recorded as many as six names, then asked, "Do you feel equally close to all these people? (IF NO): Which of these people do you feel especially close to? (PROBE: Anyone else?)" They next asked about ties among all pairs of alters: Please think about the relations between the people you just mentioned. Some of them may be total strangers in the sense that they wouldn't recognize each other if they bumped into each other on the street. Others may be especially close, as close or closer to each other as they are to you. First, think about NAME1 and NAME2. Are they total strangers? Are they especially close? (PROBE: As close or closer to each other as they are to you?)

Next, the interviewer asked how long the respondent had known each alter, how often they talked to one another on average, and various types of roles they played in relation to the respondent (e.g., spouse, parent, child, neighbor, coworker, friend, advisor). Finally, the respondent also reported each alter's gender, race, education, age, religion, and political party identification.

Figure 2.1 provides a visual illustration with an artificial example of an ego naming three alters based on the name generator question. After the nominations, the ego is asked of the relationship between it and each of its alters, such as spouse, parent, or neighbors etc. The ego is also asked of the ties between its alters. In our artificial example, ego reports that alter1 knows alter2, but none of them knows alter3. In addition, the ego is also ask about alter's information such as gender, race, age, and education etc.

As this example reveals, the ego-net survey tasks the egos with a sizable amount of work in naming their personal network, especially when the egos nominate multiple alters. If ego names no alter, it does not have to provide any more information in this line of surveys. In contrast, if ego names all six alters (reaching the upper limit), it will have substantive increase in number of questions to answer. It would have to assess its relations with all six alters and answer all the questions

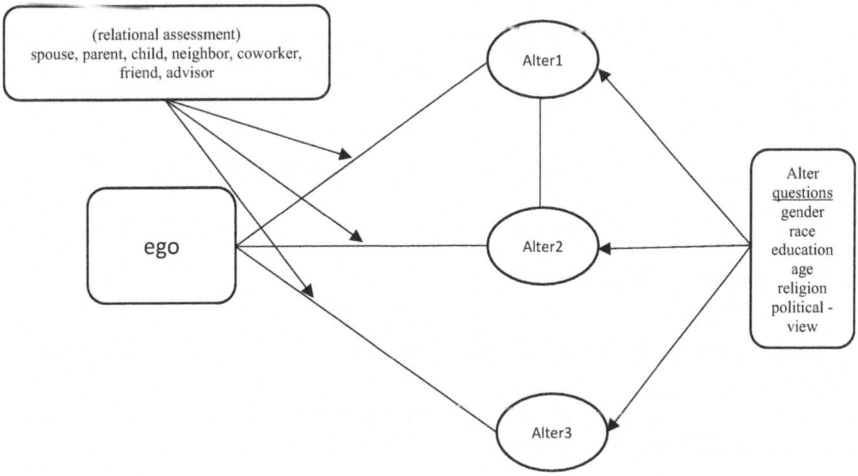

Fig. 2.1 An artificial example of an ego naming three alters in name generator

(gender, race, age, education etc.) about each of the six alters. It would also have to provide information about the 15 dyadic relations among the six alters ($C_6^2 = \frac{6!}{2! \times 4!} = 15$). In fact, the number of questions imposed on the ego to respond increase exponentially with each additional alter nominated by the ego. Such observation is one of the major concerns on McPherson et al.'s [26] finding that American is becoming increasingly lonely in the early twenty-first century compared with it 20 years ago in mid-1980s. The following sections discuss this in detail.

The GSS in 1985 used the name generator to elicit personal network from US adult population [24]. Because the name generator specifically asked respondents to name their contacts with whom they discuss important matters, the personal network elicited from the survey is also called core discussion group/network of Americans. The GSS 2004 replicated this design, conducting another round of survey of American's core discussion group. McPherson et al. [26] analyzed both panels, reporting that Americans become more isolated and less connected between 1980s and early twenty-first century, causing a great deal of debates among intellectuals, as well as general publics. Key evidence supporting the isolation argument is that the average size of core discussion group decreased from 2.94 in 1985 to 2.08 in 2004. In 1985, only 10 percent of respondents reported having no one to discuss important matters, that percentage increase to 25 percent in 2004. Meanwhile, education and racial homogeneity in core discussion network remain high in both surveys, while gender composition is very well mixed and kinship portion continues to be high. It appears not only Americans core discussion group shrinks in size, but also Americans increasingly recoil from network outreach, naming close family members in their core discussion network.

As mentioned early, the results received some doubts arising from the questionnaire anomalies [8]. Researchers cited the lack of training and persuasion on the part of interviewers, and the awareness of ego-network design on the part of interviewees caused the network size to be artificially low [30]. Another line of criticism focused on the online social media network, arguing that early twenty-first century witnessed the beginning of online social media that is increasingly prominent in the way people are connected with each other [28]. In eliciting core discussion group members with the traditional name generator, the 2004 GSS did not capture personal social media network. Nevertheless, name generator is one of the earliest and most important survey instruments used to capture individual personal network.

2.2.2 *Positional Generator*

Positional generator is another important instrument to capture individual personal network in ego-net design. Lin and Dumin [18] are credited with developing the positional generator and applying it to elicit personal network. Much like name generator, positional generator starts with a single questionnaire item,

Among your relatives, friends, or acquaintances, are there people who have the following jobs?

A. High school teacher
B. CEO of a company
C. Nurse
D. Handyman

After the respondents name their social contacts along the job titles in the survey, the interviewers continue to ask "for each job for which the respondents answer yes, ask what is his or her relationship with the respondent,

A. Relative
B. Friend
C. Acquaintances

Key in the positional generator ego-net design is the choice of the positions to ask.

Lin and Dumin [18] and Lin [21] advocated that those positions should be the most popular occupations in US Census. For example, the 2004–2005 U.S. Social Capital Survey [20] adopted such approach to identify 22 occupations that were used to gauge respondents social contacts. Those positions represent a broad occupational spectrum, from low-level manual labor to high-level managerial or professional positions. A crucial difference between positional generator and name generator lies in the underlying theoretical motivation behind each design. Name generator is used to elicit personal network on discussion of important matters, such as financial situation or personal/intimate relationship [1]. Whereas positional generator is used to gauge individual access to critical social resources to facilitate their own actions, such as finding a job via personal ties with key HR professionals in charge of hiring [2]. Positional generator is one of the most important survey instruments in ego-net studies to reveal individual's social capital [21, 35].

2.2.3 Resource Generator

While positional generator captures individual's social capital by looking at its social ties.

with others in certain positions/occupations, it builds on a premises that individuals can gain access to and mobilize the resources controlled by their network contacts in those positions. Resource generator moves one step further to tap into individual's access and mobilization of social resources by asking respondents directly whether they know people who can provide various assistances. Take a look at the following example from [34].

Do you know anybody who ......

A. Can repair a car, bike, etc.
B. Can visit socially.

C. Repairs household equipment
D. Knows about government regulations.
E. Can give a good reference for job application.

For each item the respondents answer "yes," what is his/her relation to you?

A. Family members
B. Friends
C. acquittances

Despite the differences in the probing questions, resource generator and position.

generator bear some similarities. Both tap into the resources controlled by the social contacts of respondents, and use such as proxy indicator of respondent's social capital. One empirical study [34] tested four distinctive resource generator sub-groups with 35 resource generators. One of the subgroups, dubbed as "access to prestigious occupations" corrected strongly with Lin and Dumin's [18] positional generator, while the other subgroups, such as access to information and access to instrumental supports, correlate weakly with the positional generator. This study reveals that although the two generators share some similarities, they also elicit different types of social resources.

2.3 Compare and Contrast Between Ego-Net and Whole Network Analysis

Ego-net and whole network analysis tap into different sphere of research interests and investigative topics. If the research goal is to divulge issues and topics related to the general public, then ego-net is perhaps the best way to approach the subject. The well-known GSS core discussion group studies reveal the social connectivity among American people using name generators to survey the representative samples of US population. Likewise, positional generator and resource generator disclose social resources and social capitals that can be accessed and mobilized by social actors through their social ties with others in their network. However, if the research goal is to investigate a certain specific type of social relation among a finite set of social actors, such as advise network among a group of coworkers, drug and needle sharing among a set of addicts, collaboration network among colleagues in a R & D Department, or gossip network among a cohort of graduate students. Then the whole network approach is the way to go, one key advantage of whole network analysis as opposed to the ego-net analysis is its ability to map the connections and to visualize the graph with well-developed method such as multidimensional scaling (MDS) [14]. In fact, due to the similarity between the whole network and other networks, such as computer network, telephone network, supply chain network, collaboration, and coauthor network in the data structure, many established methodologies, such as graph theory in Math and Engineering can be readily applied to analyzing social network in its whole network form [29].

Ego-net and whole network can be transitional to each other. Consider the following example that an ego starts its day with morning coffee with a group of its friends; after that it has a meeting with a few of its colleagues on research project; at around lunch time, it plays pickup games with its basketball buddies; in the evening it ends the day with a bible study session with its bible group. If we focus on each of those networks, and survey all participants of their relations with each other, we will have four different whole network matrices. Figure 2.2 shows the process.

Note that in generating the ego-net data as depicted in the Fig. 2.2, researchers would have to survey egos/respondents with four name generators, namely "with whom you have morning coffee with," "with whom you discuss research project," "with whom you exercise playing basketball," and "with whom you have bible study session." However, in creating the data entries for the four whole network matrices with the ego-net data, those name generators need to be replaced with different survey questions to tap into the relationship between the participants. For example, the five individuals in the whole social network matrix derived from ego-net's coffee drinking, so by default, everybody in this whole social network is coffee buddies with each other. Researchers need to develop an alternative survey question to tap into the finer distinction of the relationship among the five coffee buddies. Those instruments can be "with whom you feel very close to," or "with whom you discuss very important matters." In a way, the ego-net simply produces a sampling frame that researchers can use to conduct relational-specific network analysis of the whole network of those participants. Following the ego in the ego-net in Fig. 2.2, researchers can derive four whole network matrices: the coffee group, the research group, the basketball/hobby group, and the bible group.

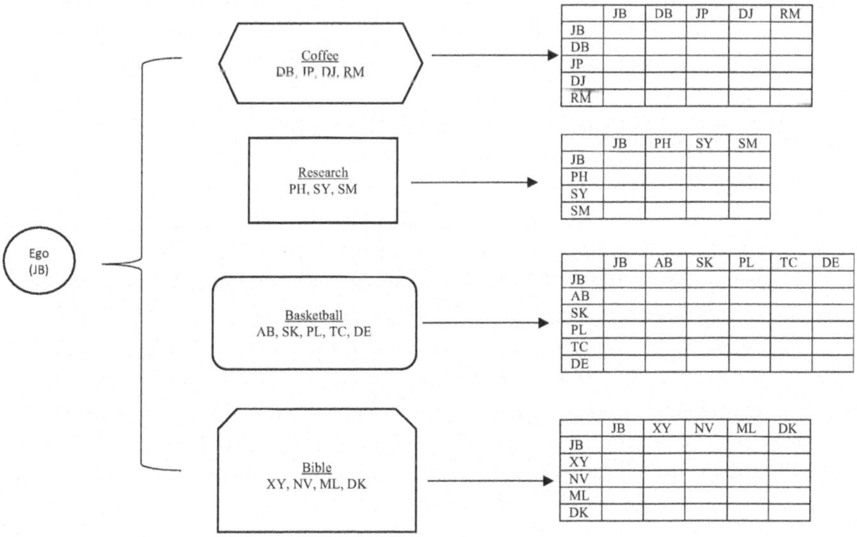

Fig. 2.2 An artificial example of generating whole network from ego-net

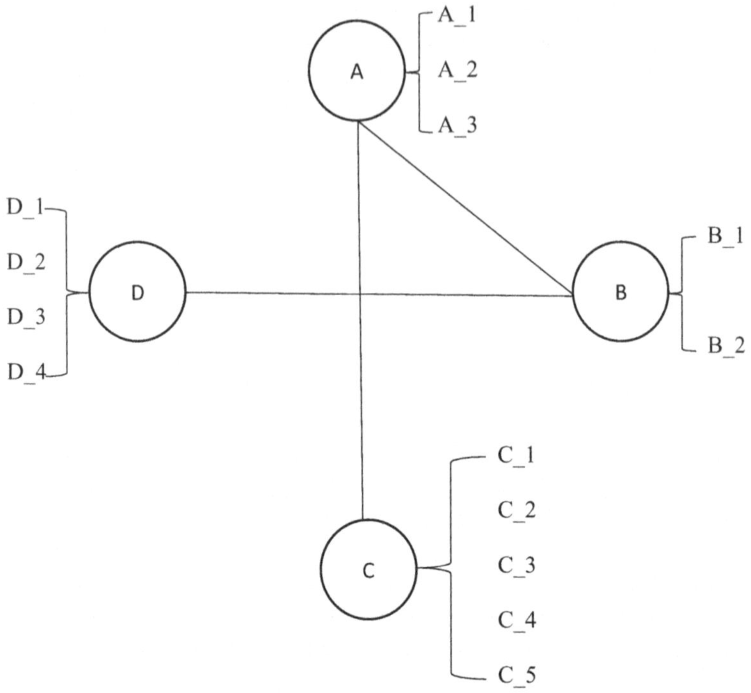

Fig. 2.3 A close friend network: from whole network to ego-net

Much like the ego-net that can be used to produce whole network data, the whole network data are also transitional to ego-net. Considering an artificial example of a small research R & D group of four scientists, a whole network approach would ask each of the scientist to identify a close friend within the group, producing the graph shown in Fig. 2.3.

In this hypothetical situation, scientist A identifies scientists B and C as its close friends, while scientist B identifies D as a close friend. For simplicity, I assume that identification of friend is mutual (if a close friendship is identified by either actor of a dyadic pair, the relation of a close friendship is present), so the graph is an undirected binary graph. Now assuming researchers ask each of four scientists to identify up to five contacts with who they discuss important matters throughout their life, it would produce four ego-net revolving around each of the scientists. In this example, one whole network data eventually produce four ego-net data, scientist A identifies 3 alters, B names 2, C produces 5, while D comes up with 4 contacts. Note that the identifications of important matter contacts does not preclude the contacts in the whole network (scientist A can name C as its discussion partner), however, it is very likely that once the nomination boundary is removed, network actors commonly name somebody outside their whole network settings.

Although the data are transitional between ego-net and whole network, researchers often do not derive the ego-net from whole network or derive whole network from ego-net. Each approach has a distinctive set of sampling strategies and associated methodologies. That said, the above discussions of data transition are crucial to understanding the advantages and disadvantages of each approach in the general paradigm of social network analysis. The implementation of ego-net is pretty easy and straightforward, simply adding name, position, or resource generators to any representative surveys in their questionnaire item lists, researchers would have the ego-net that speaks to the general population the surveys endeavor to represent. In addition, ego-net also provide researchers with a full-fledged view of ego's social network activities, revealing how individuals conduct their social activities in their social network circles for a regular day.

In contrast, the whole network could not provide a complete view of an individual's social life and social networking activities. However, the whole network data have a very different motivation and research methodologies at its disposal. Analyzing the whole network data, researchers can produce the complete birds-eye view of the structural connections among a set of finite nodes with a specific relation. Using the whole network data, Christakis and Fowler [3] illustrated that while happy people hook up with other happy people, unhappy individuals are associated with other unhappy persons, occupying peripheral positions in a happiness network. They also showed, with whole network data, that individuals commonly pursue sexual partners from their friend list and coworker contacts, because the three whole network between friendship network, sexual contact network, and coworker network tend to overlap significantly [3]. The capability to produce the bird-eye view of the complicated structural relations affords the whole network analysis the clear advantage that is unachievable in ego-net analysis. In addition, as discussed earlier, the similarity between whole network in social network and network in other disciplines (such as food network, supply chain network, and computer network) means that the developed methods in other disciplines, e.g. graph theory in STEM fields can be readily used to analyzing social network in the whole network format. Table 2.1 summarizes those pros and cons of ego-net and whole network analysis respectively.

Table 2.1 Compare and contrast between ego-net and whole network studies

	Ego-net	Whole network
Sampling frame	Representative sampling	Purposive sampling
Generalizability	Generalizable to the population to which the sample represents	Not directly generalizable, findings either cover the entire population, or the sampling distribution of the population is unknown
Inferential statistics	Parametric statistics: any conventional measures can be used, such as *Z, t, F ratio, and chi-square test*	Nonparametric statistics: inferential tests achieved through iterations of simulation.
Network activities	All network activities revolving around the ego	Relational specific network connections of all network participants
Interdisciplinary synergy	Easily implemented in other disciplines, particularly other social sciences and business studies.	Connect well with methodology development in graph theory in STEM

Table 2.1 shows five comparison points between the two social network analysis approaches. The above sections described in detail of the comparisons on the network activities and interdisciplinary synergy between the two. In terms of sampling frame, generalizability, and inferential statistics, the two approaches are also very different that demands a great level of explanation and discussions. The table highlights key differences between the two approaches in those dimensions with more to follow in the sections below that devote to whole network discussions.

2.4 Boundary Specifications

While sampling for collecting ego-net data is a no issue as they are add-on to the conventional social surveys, sampling of whole network data can be a challenge. Social network researchers must identify their network actors and the relations among this finite set of social actors. Whether it is a group of elementary students playing in after school programs, a team of coworkers socialize with each other, a unit of R & D researchers collaborate with each other on projects, or a cohort of graduate students exchange information. Researchers need to specify who are to be included in their studies, and which particular relationship they are focusing on. This is an issue of boundary specification in whole social network studies, which will be the focus of the following discussions.

2.5 Realistic and Nominalist Approach

While researchers have a variety of strategies and approaches in collecting whole network data, those strategies can be summarized with two general approaches and three implementational methods [19]. The two approaches are realistic approach and nominalist approach. In realistic approach, researchers rely on respondents to name others to be included in the study. For example, in studying a group of elementary students playing together, researchers start with a few children, asking them with whom they play with. Likewise, to study a cohort of graduate students engaging in information exchange, researchers begin with only a few students with questions such as with whom they frequently exchange important information. Essentially, realistic approach is a respondent-driven sampling process, respondents, or research subjects provide researchers with more respondents as they consider those as important and relevant to the studies. To study the US energy and health national policy formulation, Knoke and Laumann [15] started with a set of core organizations in the policy domain, asking them to name other organizations that are important in formulating, advocating, and selecting courses of action. By using this realistic approach, Knoke and Laumann were able to collect a set of core organizations in this policy network and analyze their networking activities in passing important legislature bills. The realistic approach is very effective in keeping the total number of network actors manageable and core to the relationships under study.

In nominalist approach, however, researchers impose a boundary limit to the actors and specify the relations to study at the beginning of their social network data collection. For example, to study a team of coworkers socializing with each other, social network researchers fist identify which group of workers they are focusing on. Or to examine a company of R & D scientists, researchers can start with the roster of the company, interviewing the scientists on the roster with whom they pursue research projects together. Nominalist approach bears great similarity to purposive sampling in general social survey methodology, both rely on existing sampling frame, such as company payroll list, class roster, or a participant list of a conference to identify respondents. Compared with realistic approach, nominalist approach runs risk of including social actors that are not very active in networking activities under study, producing a sparsely connected network graph.

Figure 2.4 displays a hypothetical example of realistic and nominalist approaches in identifying research scientists in collaborating with each other on projects. Using the nominalist approach, researchers use the company roster to identify the first four scientists and examine their collaboration networking. Unfortunately, none of the scientist in the company work together on projects. Instead, scientists B and C have their own collaboration network outside the company roster list. Interestingly, this example also discloses a deficiency of relying on realistic approach in identifying important network actors. Noted that the realistic approach would identify the cluster of collaboration network located at lower left portion of the picture, and through realistic approach, it identifies additional scientist to include (scientist C). However, due to isolation among the four scientists in the company list, the realistic approach stops short of identifying the collaboration network involving scientist B and his

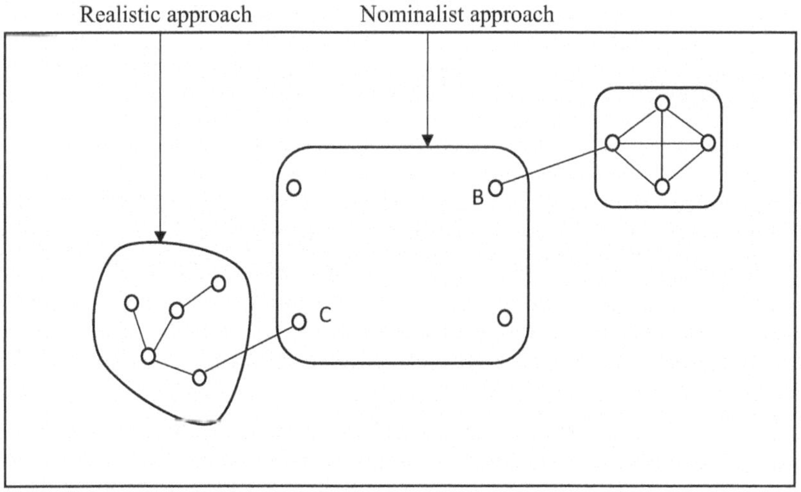

Fig. 2.4 Hypothetical example of realistic approach and nominalist approach

company in upper right corner. This hypothetical example illustrates that no approach is 100 percent perfect, one of the key challenges in researcher's report is self-critic of their network sampling with contextualizing information of their network in the general picture.

2.6 Positional, Relational, and Event-Based Methods

While the two approaches offer general guidance to social network researchers about data collection and study design, they lack details in the implementations of social network data collection. Instead, the three methods afford those detailed information that will be discussed in the following sections.

Positional method relies on the formal positions of individual actors as rule for inclusion. Those positions can be organizational membership, political party affiliation, homeowners in a residential neighborhood, or participants in some organizational entities. Empirical studies of social network analysis have been using classrooms [28], organizational employees [17], corporate board of directors [12], members of hometown association in an African village [7], or top companies (in terms of annual revenues) in Fortune 1000 and Global 500 IT industries as rule for inclusion [16].

Positional methods have a couple of concerns researchers need to be aware. One is that the inclusion rule is quite arbitrary. For example, in identifying the top companies in Fortune 1000 or Global 500, researchers can use the revenue or employment number to include the top 30 or top 50 companies in the lists. However, very often the cutoff point is set at 30 or 50 not because of the research topic or subjective interests, but because of the time and budget constraints of the researchers [10]. Perhaps very little difference exists between 31st and 30th company or between 51st and 50th company, the cutoff is arbitrary and artificial, burdening researchers with a critical task to identify potential sample selection issue or contextualize the current sample with specific issues, such as significant difference between networking of top 30 companies and partnership of top 60 companies.

Another concern of positional method is that the networking activities can be very sparse between actors in the sample. Social actors are included in the sample not because of their relationship with each other, but because they fulfill the positional criteria, having some sorts of membership, participating in a classroom, affiliated with political parties, or living in a same neighborhood. Artificially drawing a boundary for sample selections using positional criteria may end up with a collection of players not engaging many interactions, relationship building, or network connectivity. Researchers dealing with such concern should identify the networking connections not only fulfill theoretical interests, but also ensure that such connections are existent or perhaps prevalent among the sampled actors.

Reputational method relies on reputation of social actors of the social ties under study as inclusion rule. Often, reputational method starts with a set of knowledgeable informants, asking them to nominate social actors important to the relational ties of the research interests. Once those social actors are identified, researchers reach out to them with social network questionnaire surveys. Sometimes, the initial enumerated list is small and limited, researchers may need to contact those on the initial list, asking for help with identifying more actors for inclusion to the study. Michael Heaney's [13] research exemplifies the reputational method. He first compiled a list of organizations active in US Health policy lobbying. He then approached a panel of experts in academia and policy domain to elicit recommendations from the list. Those organizations that received minimum of two votes from the panel of experts are included in the study, producing a list of 171 most active organizations. Heaney then contacted those organizations about their network ties and political lobbying activities.

Reputational method is particular useful for social network sampling when available sampling frame for positional method is lacking. For example, researchers often do not have a list of participants, affiliated members, or organizational employees to sample for network analysis. Relying on knowledgeable informants, or even research subjects is perhaps the only way to generate network samples. Here, snowball sampling is another variant of reputational method that uses research subject to name additional actors to be included in the study. In this vein, snowball sampling in social network analysis is not much different from other non-social network studies. They all use existing interviewees to generate additional interviewees for the study. However, snowballing in social network analysis can be very confusing because it relies on the very ties between existing interviewees and additional interviewees for inclusion. This also causes major concerns of the representativeness of the social network samples to the sampling population. Relying on existing interviewees to produce additional interviewees can result in strongly connected social network that is highly homogenous. Despite the shortcoming, snowballing can be a powerful tool to generate a good number of respondents within a short time frame ([37]: 33). And it can be very effective identifying hidden populations, such as drug dealers, illegal immigrants, HIV communities, extreme militias, and white supremacists.

By nature of its design, reputational method is vulnerable to the acquaintance level of those knowledgeable informants for identifying respondents. Researchers must have good reasons to believe those knowledgeable informants are familiar with the targeted population and can accurately report those to include in the study, whether they are powerful members of the communities, or people of high standing in business, for example ([32]: 46). Assessing the data quality of the enumerated list from knowledgeable informants is a daunting task, which requires rigorous empirical studies. John Scott recommended that social network researchers use theoretical and empirical reasons for the choices of informants, rather than any specific social relations under investigation.

Event-based method uses certain event(s) to establish the inclusion rule, participants in those events are included in the social network analysis. Compared with the positional method or the reputational method, event-based method emphasizes on active participation rather than relying on mere membership, affiliation, or formal position with certain organizations or knowledgeable informants to nominate research subjects. Imagine this situation, researchers observe and record members of an organization who participated in "key" decision making process, using this participant roster to establish inclusion in their studies. This is what Dahl [4] used to study power structure within an organization. Rather than using formal position to proxy the political power of the agents, Dahl used the extent to which agents participate in important decision-making to determine their power structure. The famous Southern Women Study also used the event-based method to observe 14 events that involved 18 women, producing a bipartite network [6]. Compared with the positional method or reputational method, participation in the important decision-making process is more revealing of active involvement in an organization than are formal positions. And event-based method is not vulnerable to the knowledge level of those knowledgeable informants to include research subjects.

Selection of key events is critical in event-based method, that is why researchers advocate for observing multiple events rather than a single event for inclusion rule. For example, Freeman and Webster [9] observed beachgoers for 31 consecutive days and record those who attended at least three times as regulars. They then compare the data with the self-reported data collected at a later stage. Sometimes, researchers using event-based methods need to not only focus on the selection of events and their participants, but also examine those non-attendants, identifying situational force behind the missing participants. For example, business meetings rarely attract 100 percent attendance, and those missing can be for total random reasons, such as schedule conflicts, familial duties, illness etc. Occasionally, the missing attendance is reflective of disagreements or even boycott the programs intended for discussion and possible passage. In those situations, the non-attendance is actually a collective effort attempting to protest and sabotage the agendas in the meetings. Simply looking at the participants and making assumption that everybody is on board is clearly biased due to the lack of understanding of the situation by overlooking the non-participants. Again, observing multiple events and triangulate the event-based method with data from other methods can significantly reduce such biases. Summarizing the above discussions, all three methods rely on one of the key elements in drawing the boundary to sample respondents, containing advantages and disadvantages with each approach. Table 2.2 compares and contrasts the three methods on their respective strength and weakness.

2.7 Online Data Mining

Social network analysis capitalizes on three pillars, theory, methods, and applications [36]. In theory, social network analysis contains three main theoretical assumptions that critically challenge the traditional egoistic view of individual behavioral analysis, inspiring many studies that adopted the relational and structural paradigm in their analyses. In methods, social network analysis has been benefiting greatly

Table 2.2 Compare and contrast the three methods of sampling network actors

	Positional method	Reputational method	Event-based method
Key component	Relying on formal positions, membership, or affiliations to sample respondents	Relying on knowledgeable informants to identify respondents	Relying on event(s) to identify and include participants
Strength	Easy to implement, especially when employee rosters or organizational membership/affiliations are readily available.	Perhaps the only way to reach hidden populations: illegal immigrants, militia extremists, or illicit drug users	A better proxy measure of active participation or power structure than the formal position/ membership
Weakness	Boundaries delineate is arbitrary, based on researcher's time and budgetary constraints, rather than some intrinsic research interests.	Very vulnerable to the knowledge level of the key informants, may produce strongly connected network that is very homogenous	Selection of key events are crucial. Overlooking non-participants may lead to biased observations
Solutions	Start with core, gradually relaxing the boundary to include more peripheral network actors	Exercise due diligent on identifying the knowledgeable informants, making sure they are truly knowledgeable of all key players	Use multiple data source to triangulate the rosters from observing the event(s)

from its multidisciplinary roots, achieving interdisciplinary synergy that involves social sciences and science, technology, engineering, and mathematics (STEM) fields. In applications, social network analysis stimulates many studies in disciplines that traditionally build on individualistic and egoistic approaches. Outside of the three pillars, one main reason for social network analysis's popularity among academics lies in the commercialization of social networking among people in daily life. Commercial capitalization of social networking activities reflects in popular social media platforms such as Facebook, LinkedIn, Twitter, YouTube, WhatsApp, Instagram, and WeChat etc. Those platforms link billions of people, facilitating huge amount of information exchanges and social networking constantly. They provide a goldmine for social network analysis, as well as great challenge to mine, store, organize, analyze, and present these massive amounts of data that are not only massive in quantity, but also fast changing in speed. Social network scholars must overcome their comfort zones to collaborate across aisles between social scientists, computer scientist, statisticians, and others to handle the challenge of social network analysis of big data embedded in online social medias. Sections below describe the fundamental techniques and issues facing online data mining of social media data. After that, it discusses the Intelligent Cyber Argumentation System (ICAS), which was designed and implemented by an interdisciplinary team of computer engineers and social scientists at University of Arkansas through their NSF funded project Data Analytics that are Robust and Trusted (DART).

2.8 API and Web Scraping

When it comes to mining the online social media data for social network analysis, two common approaches materialize. The first one is API (Application Programming Interface), which is an interface of a computer that allows the software to speak with other software [23]. Many social media companies make their data banks available through their APIs, hence providing an opportunity for researchers to collect data on the social media users. Researchers can draft a small software script to gain access to the API to retrieve, store, and manipulate digital traces of users from those social media companies. Many of those social media companies are platforms for social networking activities, therefore, data mined from those social media platforms are social networking between their users. Hence the social media data mining and analytics and very much intertwined with the social network analysis.

Another approach to mine online social media and social networking dataset is web scraping with web crawler. Web scraping is best tool for data extraction from traditional websites such as blogs, forums, organizational pages, online newspapers, magazines, Twitters or Facebook with open access contents [33]. Depending on the technologies, and tasks at hands, web scraping ranges from very straightforward [25] to very complicated [27]. Web scraping often produces hyperlinks, and socio semantic networks that require additional technology such as Natural Language Processing (NLP) for data production.

Either through API or web scraping, the social media data generated are digital footprints of users, which are instantaneous and nonintrusive. They are the "candid shots" of users' thought process through their postings, message boards, and following/friending patterns. Think about the traditional data collection through in person, telephone, or online interviews, respondents must deliberate their answers to each question in the survey. In contrast, those social media data disclose to researchers of the spontaneous actions or reactions toward posts, messages, or social networking activities of the social media users. Data garnered from traditional social surveys and from online data mining can be drastically different, with advantages and disadvantages associated with each approach.

We must caution readers of data garnered through API regarding the validity, reliability and generalization issues. First on validity, which measures the extent to which the data represent the true contents they attempt to measure, data from online social media can contain fake profiles, and spambots. To the extent that inaccurate information is an inherent issue in social science data collection, social media data is particularly vulnerable to inaccurate information because users are under no pressure or obligations to enter accurate personal profiles. In fact, many users purposively enter inaccurate information to avoid being detected or being hacked into their true account (credit cards, or banking information etc.). In addition, online social media data are great source to reveal "what" users are doing, but relatively poor basis to disclose "why" or the reasons behind their actions. Information gathered from in-depth interviews or other close-up observations provide much richer and contextual information that greatly supplement the information from API or online data scraping [23]. Because through interviews and observations, research not only detect the communications, and interactions, but also understand the context under which those interactions and connections take place.

The next issue is data reliability, which measures to what extent the data generated from API of online social media are consistent with data generated with other modes of collection from the same source. Here, the main concern is that researchers rely on API for data production is at mercy of corporate policy and practice with API data restrictions. Companies such as Facebook and Twitter routinely restrict third party access to their datasets. Often, they made changes to their API portals without informing the researchers, creating blind spots of researchers who end up collecting partial data. Sometimes, well tested and well-developed computer codes and routines are obsolete due to the unannounced changes in API portal by the companies. Thus, to collect a more reliable dataset, we need to triangulate the API online data with data from other modes of collections, such as Computer-Assisted Personal Interviews (CAPI), personal observations, and other methods.

The third issue of API data from online social media is about data generalization, which measures to what extent findings from the online data mining can be generalized to the whole population. In general, this is a challenging topic as data from online source or social networking activities happen in online social media are not part of parametric statistics with normality assumptions. Rather they are all part of nonparametric statistics based on normality simulations, rather than assumptions.

However, in the most intuitive sense, online data, no matter how carefully researchers design the sampling process, will oversample those active content contributors and under-sample those lurkers, who are silent observers [23]. Ironically, in many online social media communities, those lurkers account for majority of the users, but they are not well represented in many social research samples. Again, researchers need to triangulate their findings by looking at both API studies of online social media and other well designed national representative samples. One great source of national sample in the field of sociology is General Social Surveys (GSS) [11] (https://gss.norc.org/). It is a high quality nationally representative sampling social surveys spanning for more than 50 years. Researchers [24, 26] have been using GSS to publish on the subject of social networking and social connections among American adults. One research area that did not touched by those studies are online social media networking, which leaves ample room for prospective studies based on API and online data mining.

2.9 Intelligent Cyber Argumentation System (ICAS)

The Intelligent Cyber Argumentation System or ICAS platform is an experimental online deliberation platform that is designed to facilitate large-scale online discussion and debate and provide integrated analytical models to summarize and analyses the discussions automatically [22, 31]. The ICAS platform facilitates online discussions that are anonymous (users are assigned a randomly generated username at registration time), asynchronous (posts are persistent and visible to users at all times), and un-moderated (all posts are immediately visible to others after posting). ICAS is an issue-centric discussion platform, meaning that all of the discussions center around a topic issue. Although the ideas of facilitating online discourse and monitoring its progression is not new [22], the actual platform of ICAS was not developed until the NSF funded DART (Data Analytics that are Robust and Trusted), which allows scientists to work together to create this online discourse platform [31].

An important note of clarification is that although the users in the chatroom are anonymous to each other, they are totally visible to us as researchers in terms of their gender, race, and other bio markers information. So, the chatroom behaviors in terms of agreeing, disagreeing, arguing, and even attacking are totally reflective of opinions, viewpoints, and posts. This is also very different from in person social networking in terms of discussing and debating because in those in-person settings, people form their discussion groups based on gender, and race. A famous sociology study on this [26] informs that people with same gender and race traits may attract each other, creating a structure with many subgroups with high racial and gender homophily exist within but high racial and gender heterogeneity between those groups.

The ICAS was deployed to collect data on online discourse to hundreds of students enrolling in a General Sociology (Introduction to Sociology) course at one South East Conference (SEC) flagship campus. The controversial social issues that were used to generate online discourse were (A) health care: should individuals be required to have health insurance? (B) same sex adoption: should same sex married couples be allowed to adopt children? (C) guns on campus: should students with a concealed carry permit be allowed to carry guns on campus? And (D) religion and medicine: should parents who believe in healing through prayer be allowed to deny medical treatment for their child? The experiments started in Fall 2017, continuing in Spring 2018, Spring 2019, and Spring of 2020. The first three rounds of data collection generate consistent numbers of participants and posts, but the last round in Spring of 2020 saw a significant decline in both participants and posts due to the pivot to online class in the middle of semester triggered by covid-19 pandemic. Table 2.3 shows the basic statistics of the ICAS data collection for the four cohorts.

Table 2.3 Participation and Post of ICAS Data Collection of the four Cohorts

Study	Total number of participants	Total number of posts
Fall 2017	318	5722
Spring 2018	335	10,573
Spring 2019	251	6384
Spring 2020	129	1476

Source: Sirrianni JW (2020) Argumentation stance polarity and intensity prediction and its application for argumentation polarization modeling and diverse social connection recommendation, Ph.D. Dissertation ProQuest 28257795, p 19

Due to Covid-19 and campus lockdown in the Fall of 2020, the research team by Douglas Adams, Joe Sirrianni, and Song Yang changed the questionnaire design and data collection using ICAS. The new questionnaire items did not ask the four controversial social issues, but instead focus on covid-related social problems, such as facemask, campus lockdown, collegiate sports, race and police/policing, and social inequality. Data collection also adopt the ego-centric design, collecting social networking around each student in our sample on those social issues. Chapter 5 of this book describes the study using ICAS data collected during Spring of 2020 to investigate networking around gun on campus discussion topic. Chapter 6 discusses the research using ICAS data collected in the Fall of 2020 on five covid-related social issues (facemask, campus shutdown, collegiate sports, race and policing, and social inequality). Note that even though both studies are social network analyses centered on social issues, study in Chap. 5 is whole network study, whereas research in Chap. 6 is ego-centric design—the beginning of this chapter discussed the critical differences between the two types of social network analysis.

References

1. Bailey SM, Marsden PV (1999) Interpretation and interview context: examining the general social survey name generator using cognitive methods. Soc Networks 21(3):287–309
2. Bian Y (1997) Bringing strong ties back in: indirect ties, network bridges, and job searches in China. Am Sociol Rev 62(3):366–385. https://doi.org/10.2307/2657311
3. Christakis N, Fowler J (2009) Connected: the surprising power of our social networks and how they shape our lives. Little Brown, New York
4. Dahl RA (1961) Who governs? Democracy and power in an American City. Yale University Press, New Haven
5. DART. https://dartproject.org/. Downloaded in May 2024
6. Davis A, Gardner BB, Gardner MR (1941) Deep south: a social anthropological study of caste and class. University of Chicago Press, Chicago
7. Feldman-Savelsberg P, Ndonko FT, Yang S (2005) How rumor begets rumor: collective memory, ethnic conflict, and reproductive rumors in Cameroon. In: Fine GA, Campion-Vincent V, Heath C (eds) Rumor Mills: the social impact of rumor and legend edited by fine. Transaction Publications, New York, pp 141–159
8. Fischer CS (2009) The 2004 GSS finding of shrunken social networks: an artifact? Am Sociol Rev 74(4). https://doi.org/10.1177/000312240907400408
9. Freeman L, Webster C (1994) Interpersonal proximity in social and cognitive space. Soc Cogn 12(3):223–247
10. Galaskiewicz J (1979) The structure of community organizational networks. Soc Forces 57(4): 1346–1364
11. General Social Surveys (GSS). https://gss.norc.org/. Downloaded in May 2024
12. Harris DA, Helfat CE (2007) The Board of Directors as a social network: a new perspective. J Manag Inq 16(3). https://doi.org/10.1177/105649260730
13. Heaney M (2014) Multiplex networks and interest group influence reputation: an exponential random graph model. Soc Networks 36(1):66–81
14. Knoke D, Yang S (2020) Social network analysis. Sage, Thousand Oaks
15. Knoke D, Laumann EO (1982) The social structure of national policy domains: an exploration of some structural hypotheses. In: Marsden PV, Lin N (eds) Social structure and network analysis. Sage, Beverly Hills, pp 255–270

16. Knoke D (2001) Changing organizations: business networks in the new political economy. Westview Press, Boulder
17. Krackhardt D, Kilduff M (1999) Whether close or far: social distance effects on perceived balance in friendship networks. J Pers Soc Psychol 76(5):770–782. https://doi.org/10.1037/0022-3514.76.5.770
18. Lin N, Dumin M (1986) Access to occupations through social ties. Soc Networks 8(4):365–385. https://doi.org/10.1016/0378-8733(86)90003-1
19. Laumann EO, Marsden PV, Prensky D (1983) The boundary specification problem in network analysis, chapter 1. In: Burt R, Minor M (eds) Applied network analysis. Sage, Beverly Hills
20. Lin N, Yang-chih F, Chen C-j J (2013) Social capital and its institutional contingency a study of the United States,China and Taiwan. Routledge, New York
21. Lin N (2001) Social capital: a theory of social structure and action. Cambridge University Press, Cambridge. ISBN 978-0-521-52167-3
22. Liu XF, Khudkhudia E, Wen L, Sajja V, Leu MC (2010) An intelligent computational argumentation system for supporting collaborative software development decision making. In: Meziane F, Vadera S (eds) Artificial intelligence applications for improved software engineering development: new prospects, pp 167–180). IGI Global. https://doi.org/10.4018/978-1-60566-758-4
23. Lomborg S, Bechmann A (2014) Using APIs for data collection on social media. Inf Soc 30(4): 256–265. https://doi.org/10.1080/01972243.2014.915276
24. Marsden PV (1987) Core discussion networks of Americans. Am Sociol Rev 52(1):122–131
25. McLevey J (2021) Doing computational social science: a practical introduction. United Kingdom, SAGE Publications
26. McPherson M, Brashears ME, Smith-Lovin L (2006) Social isolation in America: changes in Core discussion networks over two decades. Am Sociol Rev 71(3):353–375. https://doi.org/10.1177/000312240607100301
27. Mitchell Ryan E (2018) Web scraping with python: collecting more data from the modern web. Sebastopol, CA, O'Reilly Media
28. Morimoto SA, Yang S (2013) What friendship entails: an empirical analysis of graduate students' social networks. Sociol Spectr 33(2):99–116. https://doi.org/10.1080/02732173.2013.732870
29. Nooy, Wde (2009) Social network analysis, graph theoretical approaches to. In: Encyclopedia of complexity and systems science. Springer, New York, pp 8231–8245
30. Paik A, Sanchagrin K (2013) Social isolation in America: an artifact. Am Sociol Rev 78(3): 339–360. https://doi.org/10.1177/0003122413482919
31. Sirrianni J, Liu X, Adams D (2020) Agreement prediction of arguments in cyber argumentation for detecting stance polarity and intensity. In: Proceedings of the 58th annual meeting of the Association for Computational Linguistics, pp 5746–5758. https://doi.org/10.18653/v1/2020.acl-main.509
32. Scott JG (2012) Social network analysis, 3rd edn. Sage Publications, London
33. Tindall D, McLevey J, Koop-Monteiro Y, Graham A (2022) Big data, computational social science, and other recent innovations in social network analysis. Can Rev Sociol 59(2):135–288
34. Van der Gaag M, Snijders T (2004) Proposals for the measurement of individual social capital. In: Flap H, Volker B (eds) Creation and returns of social capital. A new research program. Routledge, London
35. Yang S, Jackson BA, Zajicek A (2021) Changing landscape? An intersectional analysis of race and gender disparity in access to social capital. Sociol Spectr 41(1–2):80–95
36. Yang S (2022) Social network analysis. In: Frey BB (ed) The SAGE encyclopedia of research design. Sage, Thousand Oaks, pp 1546–1551
37. Yang S, Keller FB, Zheng L (2016) Social network analysis: methods and examples. Sage, Thousand Oaks

Chapter 3
Methods for Analyzing Networks

Song Yang

Abstract This chapter starts with the basic topic of network representation with graphs and matrices. Then it introduces fundamental methods for analyzing nodes, such as centrality, power and prestige. It describes structural, regular and automorphic equivalence for analyzing dyadic pairs, and cliques for examining subgroup partitions. It concludes with discussions on measures of whole network, such as density and centralization.

Keywords Graph Theories · Social Network Representations · Social Network Analysis Methods

This chapter starts with the basic topic of network representation with graphs and matrices. Then it introduces fundamental methods for analyzing nodes, such as centrality, power and prestige. It describes structural, regular and automorphic equivalence for analyzing dyadic pairs, and cliques for examining subgroup partitions. It concludes with discussions on measures of whole network, such as density and centralization.

This chapter strives to provide a basic introduction to various analytical methods of social networks. Several points of clarification are worth mentioning before the introductions. First, those methods presume a whole social network. The previous Chapter (Chap. 2) describes the differences between ego-centric (or ego-net) social network and whole social network in terms of research design and data collections. Various analytical methods exist for analyzing ego-net, space restriction of this monograph preclude coverage of this area, for details of analyzing ego-net, please consult Crossley et al. [4] and Marsden [6].

Second, those whole social networks (or graphs) are further distinguished into four types, from the simplest to the complicated in terms of analytical methods, they are binary undirected, binary directed, valued (weighted) undirected, and valued (weighted) directed graphs. Based on my survey of the field, developments in

S. Yang (✉)
Department of Sociology and Criminology, University of Arkansas, Fayetteville, AR, USA
e-mail: yangwang@uark.edu

S. Yang (ed.), *Social Network Analysis in Action*, Lecture Notes in Social Networks,
https://doi.org/10.1007/978-3-031-66661-2_3

analyzing valued graphs are spotty and random, depending on the research issues at hands. Because this book is devoted to providing a systemic introduction to various methods in analyzing social networks, I will focus on the binary graphs (directed or undirected), leaving audiences to search for different methods in analyzing valued graphs per their specific research needs. For a clarification, Fig. 3.1 presents a visual classification of the four types of social network (for details of social network typology, please read [7]).

As we are on this topic of typology of social networks, a formal definition of social network is defined as a set of social actors and the dyadic social ties connecting those actors. Those social actors can be individuals, or collective entities such as teams, groups, parties, neighborhoods, organizations, or nations/states. Likewise, the dyadic ties (pairwise connections) can be emotional ties in terms of emotional supports, and/or instrumental ties such as tangible helps (childcare, auto or house repairs, personal loans, information sharing). Dyadic ties include (but not

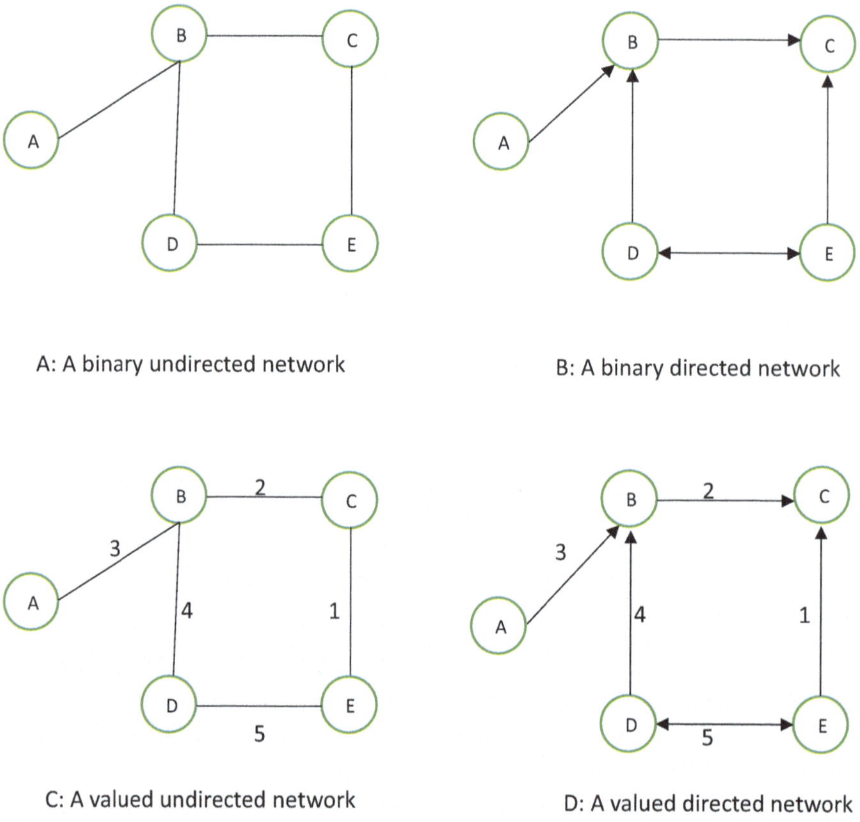

Fig. 3.1 A visual illustration of four analytical types of social network. (**A**) A binary undirected network. (**B**) A binary directed network. (**C**) A valued undirected network. (**D**) A valued directed network

limited to) friendships, communications, co-authorships, dating, sexual relationships, trade, co-conspiracy, contagions (think of Covid-19 contact tracing), etc. Noted that graph theory is also a classical research topic in STEM, which represent the graph with a succinct notation as G (V, E). In this notation, G represents graph, whereas V denotes vertices (nodes or actors in social network analysis), and E indicates edges (ties, relationship or connections in social network analysis).

Third, although developments in methods of social network analysis started with in person (face to face) social networking as analytic foci, those methods can be equally useful in analyzing social networks that take place in virtual space, such as computer or smart phone based social networking (Facebook, Twitter, LinkedIn, WeChat etc). I stress in Chap. 2 that computer supported social networking provide challenges as well as opportunities for social scientists to work across the aisle with computer (data) scientists to mine the massive online data of social networking. Granted computer supported social networking requires a different set of skillsets in data scraping, storage, processing, and even analytical skills, methods described in this chapter can be used to analyze all types of data, regardless of the modes of social networking (be it as personal face to face or online/smart phone networking).

3.1 Network Representation

Social network is represented with two main forms, graph and matrix. A graph is a visual illustration of social network among a finite set of actors. A matrix is a mathematical representation of a social network. Although graph presents a good visualization to illustrate social network to audiences, matrix is the only way that permits mathematical and computer operation of social network analysis. In particular, matrix that represents social network is commonly called adjacency matrix, and depends on the graph it tries to denote, it can be binary adjacency undirected matrix (in representing binary undirected graph), binary directed matrix (in representing binary directed graph), valued undirected matrix (in representing valued directed graph) and valued directed matrix (in representing valued directed graph). Table 3.1 displays those four different types of matrices corresponding to the four configurations in Fig. 3.1.

The matrix representation of social networks takes the squared table format, in which the number of rows equal the number of columns. In adjacency matrix, the sequence of actors in rows (A, B, C, D and E) is the same as the sequence in columns (A, B, C, D and E). The cells in the matrix indicate the presence/absence of ties between the pair of actors, which is commonly denoted as $X_{i,\ j}$, in which i is the row and j is the column ($X_{i,j} = 0/1, when\ i \neq j$). In valued graphs, the cells would indicate the values of the relational ties between a pair of actors. The diagonal values in adjacent matrix all equal 0 ($X_{i,\ i} = 0$), because actors do not have meaningful relationship with themselves. Noted also that the off-diagonal cells indicate direct connections between pairs of nodes. The matrix does not include indirect connections between all pairs in the graph, it is perhaps for this reason, those matrices are

Table 3.1 Adjacency matrices of four analytical types of social network

A: A binary undirected network

	A	B	C	D	E	RM
A	0	1	0	0	0	1
B	1	0	1	1	0	3
C	0	1	0	0	1	2
D	0	1	0	0	1	2
E	0	0	1	1	0	2
CM	1	3	2	2	2	–

B: A binary directed network

	A	B	C	D	E	RM
A	0	1	0	0	0	1
B	0	0	1	0	0	1
C	0	0	0	0	0	0
D	0	1	0	0	1	2
E	0	0	1	1	0	2
CM	0	2	2	1	1	–

C: A valued undirected network

	A	B	C	D	E
A	0	3	0	0	0
B	3	0	2	4	0
C	0	2	0	0	1
D	0	4	0	0	5
E	0	0	1	5	0

D: A valued directed network

	A	B	C	D	E
A	0	3	0	0	0
B	0	0	2	0	0
C	0	0	0	0	0
D	0	4	0	0	5
E	0	0	1	5	0

called adjacency matrix. For example, in Fig. A of Fig. 3.1, A is connected to B, which is in turn connected to C, so A and C are indirectly connected via B. However, due to this connection between A and C being indirect through B, the adjacency matrix in A of Table 3.1 shows 0 in the cell of A and C.

In undirected network, $X_{i,\ j} = X_{j,\ i}$, *when* $i \neq j$. For example, in Table A of Table 3.1, $X_{A,\ B} = X_{B,\ A} = 1$. However, in directed network, $X_{i,\ j} = / \neq X_{j,\ i}$, *when* $i \neq j$, because in directed network, senders of the relational ties (commonly denoted as *i* in rows) are distinguished to be different from the receivers of the relational ties (commonly denoted as *j* in columns). For example, in Table B of Table 3.1, $X_{A,\ B} = 1$, *whereas* $X_{B,\ A} = 0$, because in Fig. B of Fig. 3.1, A sends a tie to B, but B does not send a tie back to A.

The difference between valued/weighted graphs and binary graphs is that in valued graphs, dyadic pairs carry values to indicate the intensity of relationship between the pairs, whereas in binary graphs, dyadic pairs are different in binary version (either presence or absence of the direct ties). For example, in Table C of 3.1, actor E is connected with actors C and D at level 1 and 5 respectively, indicating different levels of relational intensity between the pairs of EC and ED. Whereas in Table A of 3.1, actor E is simply connected with actors C and D, with no value differentiations to further distinguish the intensity of the relationship. Valued graphs are distinguished between undirected and directed graphs, parallel to the distinction in binary graphs. Essentially, binary graphs, especially undirected binary graphs are easiest to analyze, while it is much more challenging to analyze valued graphs, especially directed valued graphs. Many of the analytical methods are restricted to analyzing binary graphs, and I will be careful to clarify the type of networks the analytical methods apply in discussing computational details of those methods.

3.2 Centrality

Centrality is an actor/nodal level indicator that measures the extent to which a given actor occupies the center of a social network. Centrality is perhaps one of the most important social network indicators of individual actor's position in a given network, and for that reason, it is discussed as the very first social network analysis measurement in almost all the social network analysis books. Another reason for its being discussed first is its simplicity: simply to calculate and to interpret. Individual actor's centrality in a social network is often used to denote its visibility, importance, commanding of information, or popularity, depending on the type of social networks. For example, those who are central in the friendship networks are popular among its peers/colleagues. Those central actors in information network receive larger amount of information than those in the peripheral positions. Nations occupying central positions in international trade network are highly visible and important players of the global trade network. In centrality computations, four measurements are commonly used, such as degree centrality, closeness centrality, betweenness centrality, and eigenvalue centrality. The following sections explain each centrality in detail.

3.3 Degree Centrality

Degree centrality measures the extent to which an actor is directly connected with other actors in a given network. In binary undirected graphs, the connections do not have directions, thus degree centrality of an actor is aggregation of all direct ties between the actor and other actors in a given network. Hence the following computation formula for degree centrality in binary undirected network,

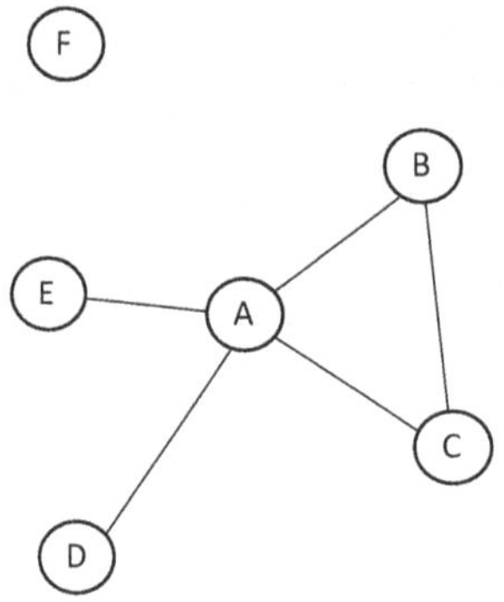

A: binary undirected graph

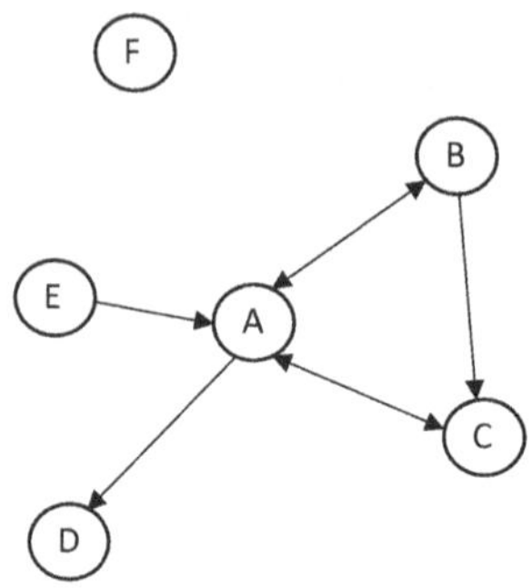

B: binary directed graph

	A	B	C	D	E	F	RM
A	0	1	1	1	1	0	4
B	1	0	1	0	0	0	2
C	1	1	0	0	0	0	2
D	1	0	0	0	0	0	1
E	1	0	0	0	0	0	1
F	0	0	0	0	0	0	0
CM	4	2	2	1	1	0	N = 10

C: Matrix representation of A graph

	A	B	C	D	E	F	RM
A	0	1	1	1	0	0	3
B	1	0	1	0	0	0	2
C	1	0	0	0	0	0	1
D	0	0	0	0	0	0	0
E	1	0	0	0	0	0	1
F	0	0	0	0	0	0	0
CM	3	1	2	1	0	0	---

D: Matrix representation of B graph

Fig. 3.2 Artificial social networks of binary undirected graphs and binary directed graphs. (**A**) Binary undirected graph. (**B**) Binary directed graph. (**C**) Matrix representation of A graph. (**D**) Matrix representation of B graph

$$DC(X_i) = \sum_{j=1}^{N} X_{i,j} \quad (i \neq j)$$

Figure 3.2 shows two artificial examples of binary undirected graph (graph A) and binary directed graph (Graph B). The Matrices below each graph demonstrate its matrix representation respectively. Degree centrality for each actor in Graph A is simply its row margins (RM) or column margins (CM) because those margins are summations of the direct connections between a given actor and all other actors in the network. For example, actor A has degree centrality of 4, indicated by its RM or CM, which makes it the most central actor in the entire network. In contrast, actor F has degree centrality of 0, indicating its lack of connection or total isolation from all other actors in the network. Noted that adding all RMs or all CMs for all actors in the network produces a number of 10, which is twice as much as the actual number of ties being present in the graph. In other words, this network has a total of 5 direct ties among its all pairs, which is an important part of calculation of network density (to be discussed in a later section).

Graph B in Fig. 3.2 shows binary directed network, with Matrix D representing this network in matrix format. In directed network, $X_{i,\,j} = l \neq X_{j,\,i}$, *when* $i \neq j$, in most cases,

$X_{i,\,j} \neq X_{j,\,i}$, in cases where $X_{i,\,j} = X_{j,\,i}$, it is more of accident than of by design. Also in the notation, $X_{i,\,j}$, i is the sender, and j is the receivers; in table/matrix format, i is the row and j is the column. For example, in Table D of Fig. 3.2, $X_{A,\,D} = 1$; whereas $X_{D,\,A} = 0$; because while A sends a tie to D, D does not send a tie back to A. In this case, $X_{A,\,D} \neq X_{D,\,A}$. In contrast, mutuality occurs between the two pairs of "A and B" and "A and C." $X_{A,\,B} = X_{B,\,A} = 1$; $X_{A,\,C} = X_{C,\,A} = 1$. However, mutuality ($X_{i,\,j} = X_{j,\,i}$) in directed graphs is exceptional, not rule.

In Matrix/Table D, RM is summation of an actor across different columns within a given row, consequentially, they represent the ties emitted by the actor to all other actors in the network. It is for this very reason, RM in binary directed Matrix is often called out-degree for the actor. In contrast. CM in binary directed Matrix is summation of an actor across different rows within a given column. So, they present the number of ties received by the actor and is often called in-degree for the actor. Looking at the Graph B or Table D, actor A sends 3 ties out to others, and receives 3 ties from others. Actor B sends 2 ties out ($RM_B = 2$) by receives 1 tie ($CM_B = 1$). In most cases in directed graphs, an actor's RM is different from its CM ($RM_i \neq CM_i$). However, in some cases, one's RM equals one's CM ($RM_i = CM_i$), such as actor A in Graph B. Those cases are exceptions, not rules.

Actor A in Graph B appears to be the most active player as both its out-degree (3) and in-degree centrality (3) are the highest among all the actors. Actor A sends out 3 relations to actors B, C, and D, and receives 3 relations from B. C, and E. Actor F is the solo isolator which does not send out any relationship, nor does it receive any from others. Actor B sends out 2 relations, receiving 1; actor C sends out 1 receives 2; actor D sends out none but receives 1; and actor E sends out 1 receives none. So, what does it mean to have different in-degree and out-degree substantively? In this vein, Knoke and Burt [5] offered one of the best substantive discussions of the differences between the two measures. In directed networks, mere participation and involvement in dyadic relationship is less important than the direction of the ties. In particular, receiving relationships, or serving as object on the receiving end of the relationship denotes power, respects, prestige, and popularity. For example, friendship nominations among a group of classmates indicate the most popular actor if the actor receives the highest number of nominations. Conversely, those who sends out many nominations, but receiving few are active players but not necessarily the popular ones.

In other types of social networks, such as information exchanges, work order and reporting, or advice seeking, receiving relationship means in reception of information, in reception of reporting, or being sought out to provide advice. In those contexts, those with high in-degree centrality command large information, receive report from its subordinates, and being popular in terms of giving advice to its colleagues. Essentially in directed networks, out-degree centrality is indicative of network activities, whereas in-degree centrality is suggestive of power, popularity, or prestige.

Degree centrality is highly susceptible to the size of social network. For example, an actor (Actor A) that is connected with 10 other actors in a social network of totally 15 members is very different than an actor (Actor B) that is connected with the same number of 10 other actors in a much larger network, say, 1000 actors. Thus, some adjustments are needed before using degree centrality to compare social actors across different networks. One common solution is to adjust for the network size by dividing actor's degree centrality by its network size. By formula, it is

$$DC = \frac{DC(N_i)}{g - 1}$$

in which DC is degree centrality, g is the total number of nodes in the network.

Such formula is called standard degree centrality. By applying it to the previous example, Actor A would have 71.4% (10/14 = 0.714) in its standard degree centrality, whereas Actor B has only 1% (10/999 = .01) in its degree centrality. Standard degree centrality reveals Actor A is very well connected to its network members, whereas Actor B is poorly connected with its networks. The standard degree centrality can surely apply to computing degree centrality for directed graphs, in which case, simply dividing the in-degree or out-degree centrality by the total number of network actors minus 1 (g−1).

Despite its simplicity, degree centrality has its limitations. One of the most obvious flaws is that it is unable to capture the situation force by overlocking the nodes with which a given actor is connected. For example, a node that is connected with another node that itself is very well connected is better connected than a node that is connected with another node that is not well connected with others. Another measure of centrality, eigenvalue centrality was developed to address such deficiency in degree centrality. Basically, to examine a node *i*'s centrality, eigenvalue centrality looks at the connectivity of other nodes (*j, k, h*) with which node *i* is connected. If those other nodes with which node *i* is connected are very well connected themselves, eigenvalue measure will be high for node *i*. Conversely, the eigenvalue will be low for node *i*.

Figure 3.3 illustrates such key difference between degree centrality and eigenvalue centrality. Actors *i* and *h* both have only one degree centrality, however, actor *i* is connected with actor A, which in turns has only one connection, whereas actor *h* is connected with actor B that is connected with five other actors. By taking account of the connectivity of those nodes with which a given actor is connected with, eigenvalue centrality serves as an important complementary measure of the connectivity of a node to other network members in a given network.

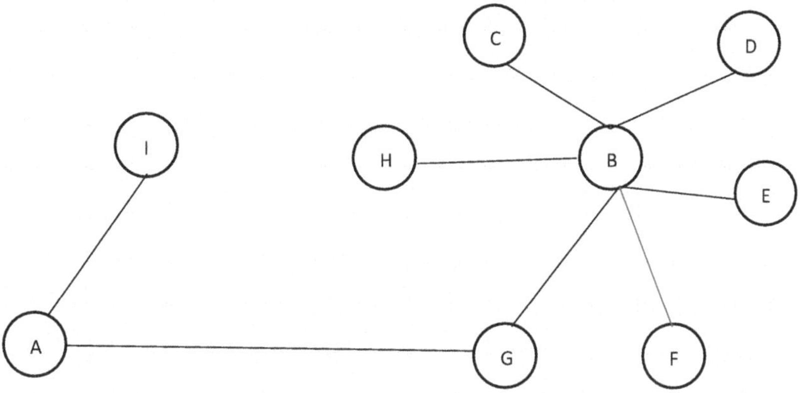

Fig. 3.3 An illustration of key difference between degree centrality and eigenvalue centrality

3.4 Betweenness Centrality

Betweenness centrality measures the extent to which a given node sits on a geodesic path between other pairs of nodes in each social network. It builds on the idea of structural hole, advocated by Ron Burt [3] that emphasizes the structural leverage of an actor connecting with other actors that are otherwise disconnected from each other. Social actors occupying many structural hole positions in a network receive two types of benefits: informational and control benefits. Empirical studies support the alleged benefits due to structural holes, as individuals with many structural holes receive large bonus and early promotion [1]; companies with many structural holes reap high profitability [2].

The computation of betweenness centrality reflects the underlying logic to capture the extent to which individual nodes position themselves in the paths connecting other pairs. In particular, assuming $g_{j,\,k}$ as the number of geodesic paths of all other pairs in a network and $g_{j,\,k}(N_i)$ as the number of geodesic paths of all other pairs that involves node i. Then, betweenness centrality has two measures, the first one is simply tallying the total number of geodesic paths of other pairs in a network that involves node i. The second one is dividing $g_{j,\,k}(N_i)$ by $g_{j,\,k}$, which reveals the extent to which node i is involved in the geodesic paths of all other pairs in a given network. Hence the following formula,

$$\mathrm{BCA}(N_i) = \sum_{i=1}^{N} g_{j,k}(N_i) \qquad (i \neq j \neq k)$$

$$\mathrm{BCP}\ (N_i) = \frac{g_{j,k}(N_i)}{g_{j,k}} \qquad (i \neq j \neq k)$$

The first formula of BCA denotes betweenness centrality in aggregation, measuring the number of geodesic paths of all pairs that involves node i. The second formula of

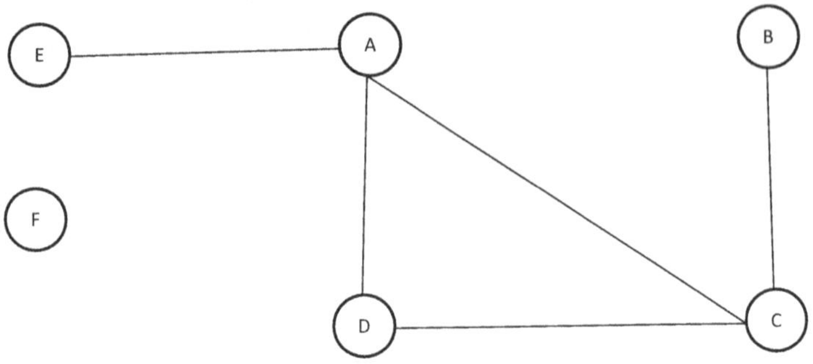

Nodes	Degree centrality	BCA	Geodesic paths	BCP
A	3	3	E-A-C-B; E-A-C; E-A-D	3/15
B	1	0	---	0
C	3	3	A-C-B; D-C-B; E-A-C-B	3/15
D	2	0	---	0
E	1	0	---	0
F	0	0	---	0

Fig. 3.4 An illustration of betweenness centrality in comparison with degree centrality

BCP denotes betweenness centrality in proportion, measuring the extent to which the number of geodesic paths of all pairs that involves node i among all pairs in the network. It ranges from 0, meaning the given node i sits on no geodesic paths of all other pairs in a given network to 1, meaning the node i sits on every geodesic path of all other pairs in a network. Figure 3.4 shows an artificial network of binary undirected graph, with the three measures (degree centrality, BCA and BCP) of the graph displayed in the table below. In all three measures, Nodes A and C are the highest. However, other nodes (B, D, E, and F) are all zero in BCA and BCP, whereas only node F is zero in degree centrality. Such exercise illustrates that betweenness centrality measures the positional leverage differently than does degree centrality. In contrast to degree centrality whereby connections with other nodes in a social network count, those connections in computing betweenness centrality do not count unless such connection positions the node in the geodesic path of other pairs.

Noted that the computation presumes that each pair has one unique geodesic path, which may not be the case in some empirical examples. When multiple geodesic paths are present, the algorithm picks one path and stays with such choice in computing betweenness centrality. In addition, the computation formulas of betweenness centrality can certainly apply to binary directed graphs. In computing betweenness centrality in those directed networks, additional cares are needed to account for direction and reachability of the geodesic paths. However, betweenness centrality may not be able to apply directly to valued graphs; alternative plan is to dichotomize those graphs into binary network before applying the formulas to computing betweenness centrality for the network nodes.

3.5 Closeness Centrality

While degree centrality measures direct connections, betweenness centrality measures strategic positions occupying structural holes in a network, closeness centrality measures the speed from one given node in a network to all other nodes. Because of this unique feature of measuring speed, closeness centrality is often used in communication network or computer network to gauge the individual node's connection speed with other nodes in the network. The computation of closeness centrality reflects the intended measurement, taking the inverse of the geodesic distance between node i and the remaining nodes in a given network,

$$CC(N_i) = \frac{1}{\left[\sum_{j=1}^{g} d(N_i, N_j)\right]} \qquad (i \neq j)$$

To illustrate the application of closeness centrality in an empirical graph, I conjure up a communication network among 7 nodes. Figure 3.5 shows the network, with its

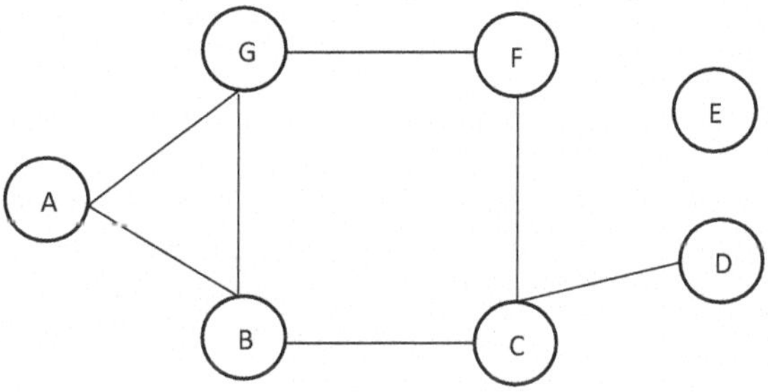

	A	B	C	D	E	F	G
A	0	1	0	0	0	0	1
B	1	0	1	0	0	0	1
C	0	1	0	1	0	1	0
D	0	0	1	0	0	0	0
E	0	0	0	0	0	0	0
F	0	0	1	0	0	0	1
G	1	1	0	0	0	1	0

Fig. 3.5 An illustration of closeness centrality in an undirected binary graph

adjacency matrix right below the graph. Considering node A's geodesic path to the remaining nodes of B, C, D, E, F, and G. Those geodesic paths from A to B would be 1, to C would be 2, to D would be 3, to E would be undefined (in calculation, the value would be 0), to F would be 2, and to G would be 1. Summing up those numbers produces the total geodesic paths of A to other nodes as 9, which computes 1/9 in A's closeness centrality. Using the same process, one can easily computes closeness centrality for node B to be 1/7. What is the closeness centrality for node D?

A few notes worth mentioning in interpreting closeness centrality. First, the computation of closeness centrality dictates that the larger the value of measurement, the closer the node is to other nodes in a network. In the above example, node B is closer to other nodes than is node A because its closeness centrality is 1/7, which is larger than that is for A (1/9). Second, the denominator in the computation of closeness centrality dictates that totally isolated nodes in a given graph do not have a valid closeness centrality measure because the isolated nodes have no geodesic paths to other nodes, which means their denominator is set to 0 (a mathematically undefined situation). Therefore, node E in Fig. 3.5 does not have a valid measure in its closeness centrality. Third, even though we use undirected binary graphs as example, the closeness centrality can be easily used in directed binary graphs. In directed graphs, special cares are needed to account for the issue of reachability from one node to another due to the directions in those graphs.

With three main centrality measures (degree, betweenness, and closeness centrality) at disposal, researchers can leverage on all three measures to triangulate network centrality for nodes in a given social network. Because the three measures emphasize different aspects of network activities, their measurements may with converge or diverge, depending on the positions and locations of each node in social networks. To illustrate, I conjure up two network graphs, one showing the convergence of the three centrality measures, and the other showing some divergence of the measures.

Figure 3.6 shows two different network graphs. In graph A on the left, node A occupies the central position in this star shape network. Node A's degree centrality is 4 via its direct ties with B, C, D, and E. Its betweenness centrality is 6, resulting from its brokerage positions in the geodesic paths of B-A-C, B-A-D, B-A-E, C-A-D, C-A-E, and D-A-E. Its closeness centrality is ¼, computing with the inverse of its direct ties with the other four nodes. In all three measures, node A has the highest centrality, indicating that it is the highest in terms of (1) direct connections, (2) strategic positioning via occupying structural holes, and (3) its speed reaching other nodes.

In graph B on the left, however, A's degree centrality is very different from its closeness and betweenness centrality. Its degree centrality is only 2 due to its direct ties with E, and F. Nodes B, C, D, E, and F all have higher degree centrality (DC = 3) than does A. However, in terms of betweenness centrality and closeness centrality, A has 12 and $\frac{1}{12}$ respectively. A's betweenness centrality results from B-E-A-F, B-E-A-F-G, B-E-A-F-H, C-E-A-F, C-E-A-F-G, C-E-A-F-H, D-E-A-F, D-E-A-F-G, D-E-A-F-H, E-A-F, B-E-A-F-G, and E-A-F-H, which sums up to totally 12 geodesic paths connecting other pairs in the network in which node A sits in

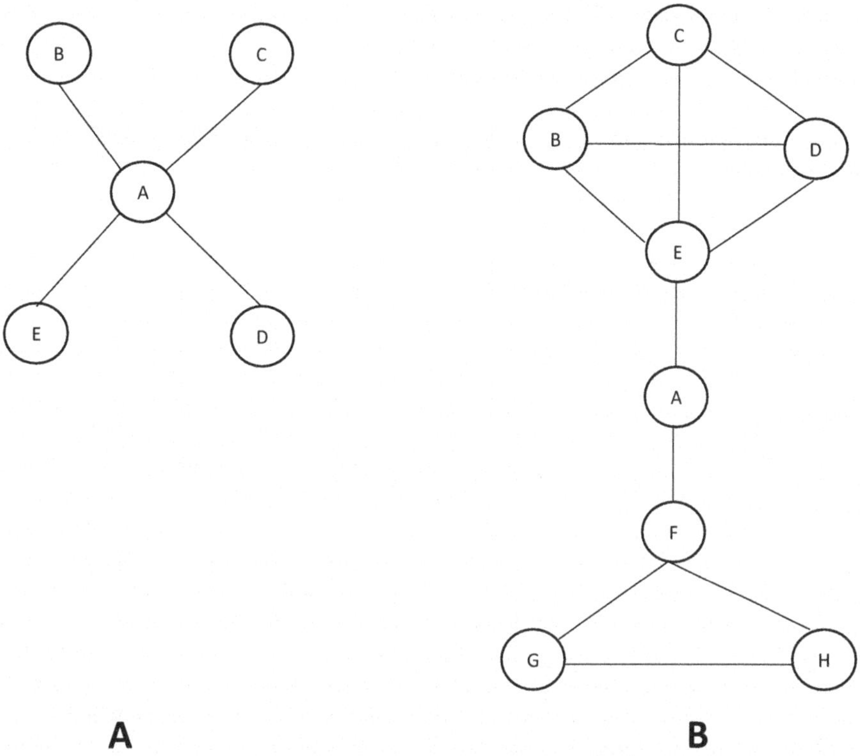

Fig. 3.6 Comparing three centrality measures with two artificial network

between. Readers can exercise computing A's closeness centrality, and also measures of betweenness and closeness centrality for all other nodes in the network. Basically, node A has the highest betweenness and closeness centrality, whereas its degree centrality is lower than many other nodes. This is an example illustrating that in many empirical networks, divergence can occur between nodal centrality measures. So, is A the central player in the network? The answer depends on which way you look at. A is not a central player in terms of direct connections with others but is a central node in terms of strategic occupying brokerage positions and having the quickest paths to others in the network.

3.6 Measures of Dyadic Relations

While centrality is about network measure of individual node in a graph, equivalence is developed and applied to measuring dyadic relations between two nodes in a network. Equivalence measures include three types: structural equivalence, automorphic equivalence, and regular equivalence. When it comes to measuring relations between pairs of nodes, one can easily think of presence/absence of ties, types of relations, or intensity of the relations. While all of those are relevant, the equivalence measures look at the patterns of their connections to other nodes between a pair of network actors, rather than the direct ties between the pair. The following sections discuss in detail of those equivalence measures, starting with the most stringent type: structural equivalence.

3.7 Structural Equivalence

Two nodes in a network are structurally equivalent if they have identical connections with all other nodes in the network. Examples of structural equivalence include two students sharing the same academic advisor, two farmers supply the produces to the same retailer, or two vendors serve as the part suppliers to the same manufacturer. Key in the structural equivalence relationship is the substitutability of the pair of nodes. Two nodes in structural equivalent relationship are substitutable with each other. If one student drops out of the advisor-advisee relationship, the network stays intact as the professor still retains the other student as its advisee. Likewise, if one farmer or part supplier were to withdraw from the network, the retailer still receives produces from the other farmer, and the manufacturer still receive parts from the other supplier. It is because of this substitutability, pairs of nodes in structural equivalence are often engage in competitive rather than cohesive relationship.

One point worth mentioning here, which is often missing in the relevant literature is the direct ties between the structural equivalence pairs. I believe when direct ties are present between the pairs in structural equivalence relationship, the competition of the pair is alleviated. However, if a direct tie between a pair of nodes in structural equivalence is absent, the competition between the pair tends to be high. Thank about the two farmers furnishing produces to the same retailer. If they have direct ties to communicate and to co-develop plans in their respective relationship to the retailer, their competition with each other to gain the retailer's favor is significant reduced. The famous Organization of the Petroleum Exporting Countries (OPEC) illustrates such co-operate strategy to manage competitive relations between oil producing nations. In contrast, if such direct connection between pairs of nodes is absent, fierce competition ensues as any node in the coalition would try hard to gain upper hands without joining forces to manage the relationship.

Definition and computation of structural equivalence between dyadic pairs depend on type of social network. In binary undirected graphs, a pair of actors i and j are structurally equivalent if, for all other nodes k in the graph, i has a tie with k if and only if j has a tie with k. Its computation formula is in the following, which stipulates that nodes i and j would be structural equivalence if $d_{ij}= 0$. Otherwise, the dyadic pair of i and j would not be structurally equivalent.

$$d_{ij} = \sqrt{\sum_{k=1}^{g-2} \left(x_{ik} - x_{jk}\right)^2} \quad (i \neq j \neq k)$$

In binary directed graphs, a pair of actors *i* and *j* are structurally equivalent if, for all other nodes *k* in the graph, *i* sends a tie to *k* if and only if *j* also sends a tie to *k*, and *i* receives a tie to *k* if and only if *j* receives a tie to *k*. Its computation formula is in the following, which stipulates that nodes *i* and *j* would be structural equivalence if $d_{ij} = 0$. Otherwise, the dyadic pair of *i* and *j* would not be structurally equivalent. Its computation formula is in the following, with the same interpretation of the result as those in binary undirected graphs.

$$d_{i,j} = \sqrt{\sum_{k=1}^{g-1} \left[\left(x_{ik} - x_{jk}\right)^2 + \left(x_{ki} - x_{kj}\right)^2\right]} \qquad (i \neq j \neq k)$$

To illustrate, I conjure a couple of graphs with one being binary undirected on the left and another being binary directed on the right, which are shown in Fig. 3.7.

The graph and tables are the left side are about an artificial example of binary undirected network. The first table right below the graph is the binary adjacency matrix of the undirected network. The second table below the matrix is the structural equivalence matrix of all pairs in the network. Noted that in undirected graph of four nodes, totally six pairs emerge: AB, AC, AD, BC, BD, and CD. Applying the computation formula for the binary undirected graph to this example produces the structural equivalence measures for all six pairs, which are displayed in the structural equivalence matrix. Out of the six pairs, two of those (AC and BD) are structural equivalent, which has 0 in their respective cell of A and C, and B and D in the structural equivalence matrix. Other four pairs are not structurally equivalent, which have large than zero values in their cells: AB, AD, BC, and CD. Much like the binary adjacency matrix of the undirected graphs, the structural equivalence matrix is also symmetric, meaning that $X_{i,\,j} = X_{j,\,i}$. For example, structural equivalence value between A and C is 0 ($X_{A,\,C} = 0$), which is the same as the value between C and A ($X_{C,\,A} = 0$).

The artificial graph on the right depicts binary directed network involving four actors. Interestingly, although the original binary adjacency matrix for directed graph is asymmetric, meaning $X_{i,\,j} \neq X_{j,\,i}$. Its derived structural equivalence matrix is symmetric ($X_{i,\,j} = X_{j,\,i}$) due to the computation formula for structural equivalence for dyads in directed graphs. Two pairs out of the totally six pairs are structurally equivalence (AC and BD pairs), as their cell values in the structural equivalence matrix are 0. All other pairs (AB, AD, BC, and CD) have 1 in their structural equivalence matrix cells, suggesting that they are not structurally equivalent.

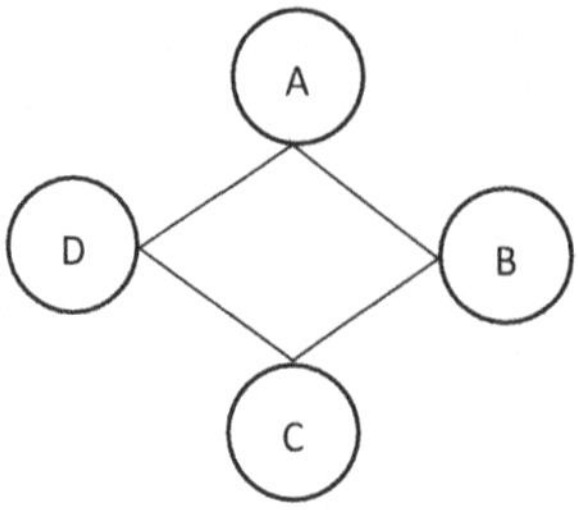

Binary Adjacency matrix

	A	B	C	D
A	0	1	0	1
B	1	0	1	0
C	0	1	0	1
D	1	0	1	0

Structural equivalence matrix

	A	B	C	D
A	0	$\sqrt{2}$	0	$\sqrt{2}$
B	$\sqrt{2}$	0	$\sqrt{2}$	0
C	0	$\sqrt{2}$	0	$\sqrt{2}$
D	$\sqrt{2}$	0	$\sqrt{2}$	0

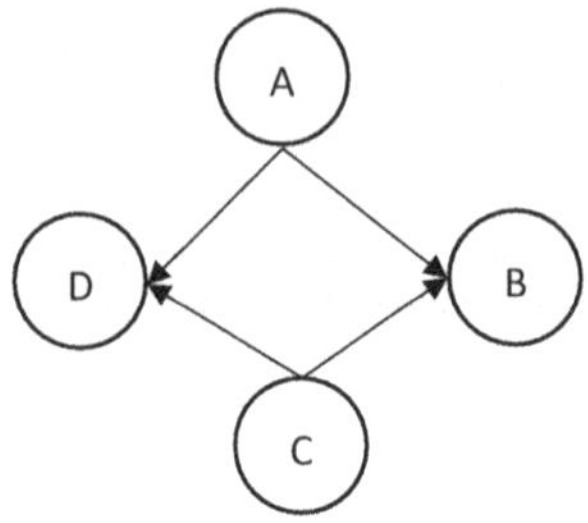

Binary Adjacency matrix

	A	B	C	D
A	0	1	0	1
B	0	0	0	0
C	0	1	0	1
D	0	0	0	0

Structural equivalence matrix

	A	B	C	D
A	0	1	0	1
B	1	0	1	0
C	0	1	0	1
D	1	0	1	0

Fig. 3.7 computing structural equivalence for pairs in binary directed and undirected graphs

3.8 Automorphic Equivalence

In measuring roles and positions in a networked society, structural equivalence is the most stringent due to its definition of two actors being connected with identical other actors in the identical ways in a given network. Automorphic equivalence and regular equivalence are two measures of dyadic relations in a more relaxing way than does the structural equivalence. For example, for structural equivalence, two managers must manage the same set of subordinates to be structurally equivalent. For automorphic equivalence, two managers can manage different sets of employees being in the automorphic equivalence relations. Figure 3.8 illustrates the difference between the two types of equivalence measures. The graph on the left depicts the structural equivalence between four actors, two of whom are managers, and two are employees. The graph on the right describes the same set of actors but in automorphic equivalence relations.

In the structural equivalence graph, two managers (managers X and Y) are managing the same set of employees of A and B. Likewise, the two employees (workers A and B) who are managed by both managers are also structurally

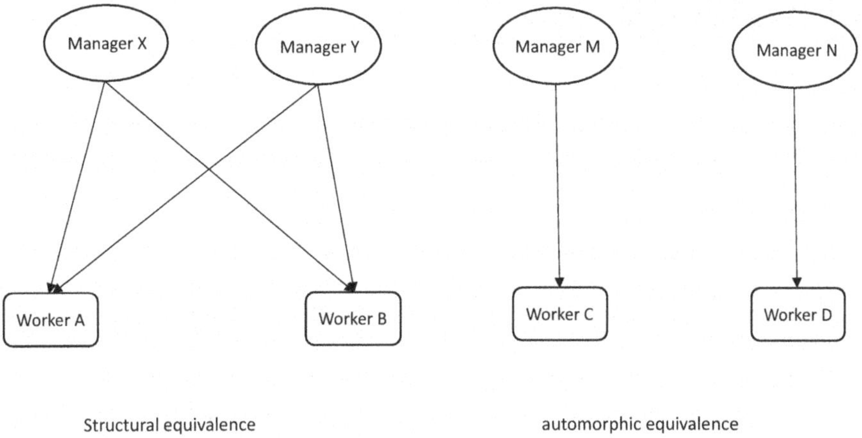

Fig. 3.8 compare and contrast between structural equivalence and automorphic equivalence

equivalent. In the graph of automorphic equivalence, the two managers (managers M and N) do not manage the same employees, but still in automorphic equivalence because they are in managerial positions with one employee under their management: manager M supervises C and manage N supervises D. Similarly, the two employees (workers C and D) are in automorphic equivalence because they each is managed by one manager, although their managers are different. From the graphs, one can easily see that structural equivalence ensures automorphic equivalence, while the reverse is not true: automorphic equivalence does not equal structural equivalence. Managers X and Y are both structurally equivalent and automorphic equivalent, and the same goes to the two employees (workers A and B), who are both structurally equivalent and automorphic equivalent. Conversely, managers M and N are only automorphic equivalent but not structural equivalent, and the same goes to workers C and D, who are automorphic equivalent but not structural equivalent.

Key feature in automorphic equivalence is that social actors in automorphic equivalence occupy indistinguishable positions. Consider that four actors in the automorphic equivalence graph in Fig. 3.8, exchanging labels between M and N, or between C and D does not alter the basic role structure or positions of the network graph. The positions between M and N and between C and D are interchangeable, thus the managerial positions between M and N, and employee positions between C and D are indistinguishable. Substantively, this feature of indistinguishable positions is crucial in understanding the role expectations and role performance of social actors: professors and students, doctors and patients, lawyers and clients. By relaxing the stringent requirement of structural equivalence, automorphic equivalence is still capable of capturing the important role expectations and performance of social actors, providing an important measure of network relations between pairs in social networks.

3.9 Regular Equivalence

Regular equivalence is perhaps the least restrictive form of dyadic relational measures in terms of social role comparisons. While structural equivalent pairs have identical ties with identical other network actors, and automorphic equivalent pairs need to have indistinguishable positions so that the two subgraphs can map over each other completely, regular equivalence does not require either condition. Rather, it depicts general role and position in relational terms. All mothers with children, doctors with patients, teachers with students, wives with husbands, or lawyers with clients are in regular equivalence relations. To illustrate the difference between automorphic and regular equivalence, I conjure up Fig. 3.9, which is shown below.

Figure 3.9 shows two pairs of lawyers (A and B, and C and D), representing 7 clients. The pair on the left between A and B are in regular equivalence positions, whereas the pair on the right between C and D are in automorphic equivalence positions. Key difference between the two pairs in equivalence measures is whether the positions are indistinguishable. In automorphic equivalence, positions between lawyers C and D are indistinguishable because they have identical network configurations (both connecting to two clients). In contrast, the pair between A and B do not share the same network configurations. A advises two clients whereas B only has one customer. However, the pair of A and B are in regular equivalence because they are both lawyers with client(s). One can easily see that regular equivalence depicts the most generalized role and positions in social network graphs, compared with automorphic and structural equivalence. Noted that neither graph in Fig. 3.9 is structural equivalence, which represents the most rigid form of equivalence between a pair of actors.

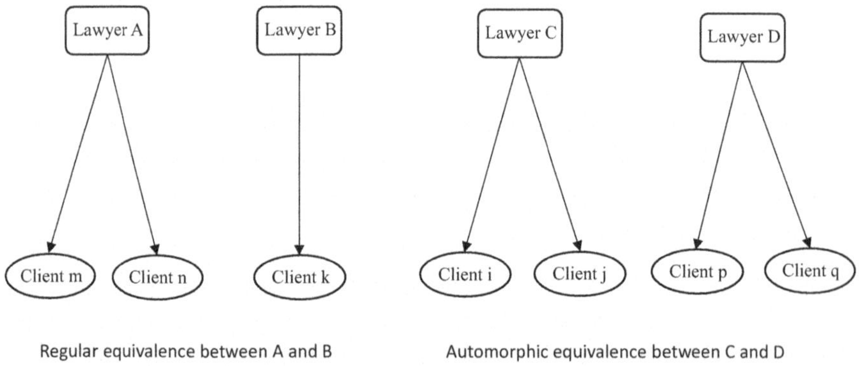

Fig. 3.9 compare and contrast between regular equivalence and automorphic equivalence

3.10 Subgroups in Social Networks

Measures of subgroups are critical part of social network analysis, as subgroups are prevalent in real life situations: think of a group of classmates being close friends with each other, coworkers collaborating and supporting each other, companies engaging in close partnership, or nation states forming close pact to combat the same axis of enemies. In the following sections, I will discuss the definitions and detecting of those subgroups, such as cliques and k-plexes. Bear in mind that intellectual fascination with subgroups goes beyond the mere detection of those, many Sociological analyses of subgroups focus on antecedent factors conducive to strongly connected subgroups such as cliques, looking at homophily along the gender, racial and social class dimensions.

3.10.1 Cliques

In many social networks, subset of network nodes can be connected via dense and enclosed ties, such as a group of friends, coworkers, and business partners. In social network analysis, researchers use an analytical tool called clique to capture and analyze those subsets or subgroup. A clique is a maximal subset of nodes ($N \geq 3$) in a social network such that every actor of the subset if directly connected with every other node in the subset. Noted that for a subset to be a clique, it needs to have at least three actors in the set. In addition, the word maximal means no other nodes exist in the network that can be added to the subset while preserving the property that every node is directly connected with each other. For example, a subset of four nodes can form a clique when each of the four nodes is directly connected with the other three nodes in the subset and no other node outside of the subset is present in the network that can be added to the subset to make it a subset of five nodes directly connected with each other. Two important characteristics are present to define and identify a clique. One is strong intra-connections with each other within the clique, and another is exclusivity of those connections to the outside members within a given social network.

Despite the usefulness of the concept clique in measuring the strong connected subgroup, many subsets in empirical observations approach near clique structure but never completely materialize into a clique structure. To capture those near clique subgroups, researchers develop another important measure called $k - plex$. A $k - plex$ of size n is a maximal subset of nodes within a network such that each node is connected to at least $n - k$ of the others. For example, a A $1 - plex$ of size 4 is a maximal subset of 4-node structure such that each node is connected to at least 3 ($4 - 1 = 3$) other nodes in the subgroup, which is actually a clique. Another example, A $2 - plex$ of size 5 is a maximal subset of 5-node structure such that each node is connected to at least 3 ($5 - 2 = 3$) other nodes in the subgroup, which is a near clique (but not a clique).

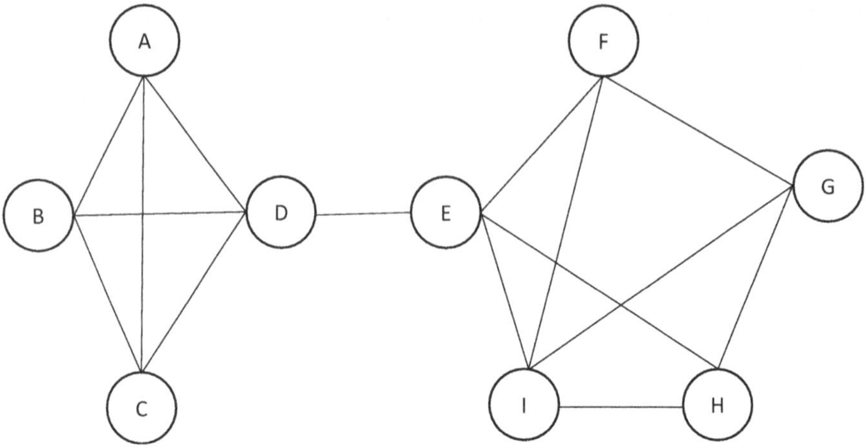

Fig. 3.10 Cliques and k-plexes in an artificial social network

To illustrate, I conjure up an artificial network of binary undirected graph that is shown in Fig. 3.10. Two strongly connected subgroups emerge in this graph, ABCD and EFGHI. The subgroup of ABCD is a clique as each of the node in this subgroup is directedly connected with other three nodes and no other node can be added to this subset without destroying the property that every node is directly connected with each other. However, the subset of EFGHI is not a clique due to lack of direct ties between E and G, and between F and H. However, the subset of EFGHI is a 2-plex because each and every node in this subset is connected to at least 3 ($5 - 2 = 3$) other nodes within this subset. Can you identify the other clique and 2-plex within this total network?

3.11 Measuring the Whole Network: Density and Centralization

Intra social network measures such as degree centrality, structural equivalence, and subgroups are important to reveal network features within a given social network. However, at the most aggregated level, the whole network deserves to be analyzed in their own rights. For example, is a whole network well connected between their nodes or are those nodes isolated from each other? Or is the whole network highly concentrated on a few opinion leaders or is it well balanced in terms of having egalitarian structures? In reality, a R & D team with well-connected communications among its scientists is conducive to many patents and innovations, whereas in contrast, a department with fragmentations and isolations would be unable to achieve high performance outcomes. Social network analysis has developed measures for whole network features. Here, I focus on two critical whole network measures: density and centralization.

3.11.1 Density

Density is an important measure of network connectivity among its nodes. It measures the connectivity by dividing the observed (actual) dyadic connections between all pairs of nodes by the mathematically maximum number of dyadic pairs between all nodes. At present, most density measures apply to binary graphs, either directed or undirected. Density calculations and interpretations of valued (weighted) graphs are more complicated than the scope of this book, we leave that topic to other special manuscripts to handle. The basic formula for computing density of binary graphs is displayed below,

$$density = \frac{total\ number\ of\ dyadic\ ties\ present\ in\ the\ graph}{total\ number\ of\ maximum\ possible\ dyadic\ ties}$$

For binary directed and undirected graphs, the detailed computation formulas for density are listed below,

$$Density = \frac{\sum\sum X_{ij}}{P_N^2} \qquad (directed\ graphs, N\ is\ the\ total\ number\ of\ nodes)$$

$$Density = \frac{\sum\sum X_{i,j}}{C_N^2} \qquad (undirected\ graphs, N\ is\ the\ total\ number\ of\ nodes)$$

To illustrate, I conjure up two artificial examples shown in Fig. 3.11. The graph on the left is binary directed, whereas the graph on the right is binary undirected. To calculate the density for the binary directed (the one on the left), we need to first count the direct ties being present in the graph, which is 6. Noticed that the ties between A and E needs to be counted twice as it is reciprocated between the two nodes. Second, we need to calculate the total mathematically maximum number of directed ties between all pairs. For this particular graph with 5 nodes, its maximum dyadic pairs are 20 pairs, resulting from $P_5^2 = \frac{5!}{(5-3)!} = 20$. Thus, the density for this graph is 30 percent (6/20 = 30%). For the graph on the right, it is a binary undirected network. Following the same process, the numerator is the observed number of dyadic ties between pairs, which is 5. The denominator is the maximum number of dyadic pairs between the 5 nodes, which is 10, resulting from $C_5^2 = \frac{5!}{2!(5-3)!} = 10$. Thus, the density for this graph is 50 percent (5/10 = 50%).

3.11.2 Centralization

Centralization is often mistaken for centrality as these two terms look very similar. In reality, they are very different as centralization measures the entire network/graph and centrality measures individual nodes in a network. Centralization indicates the

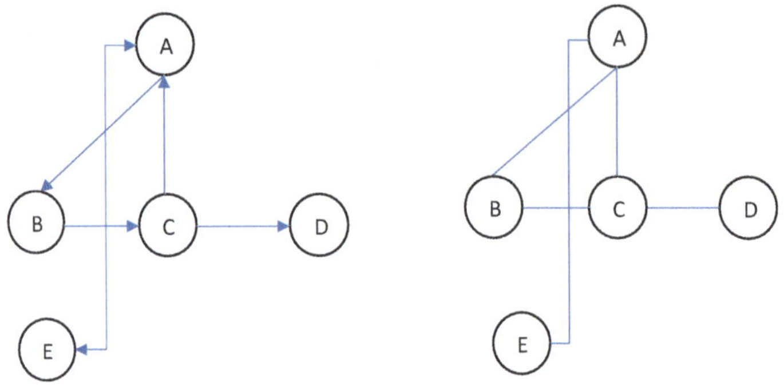

Fig. 3.11 Density in binary directed and binary undirected graphs

extent to which nodal degree centrality is different from each other in a given network. Its computation relies on the individual nodal degree centrality, but the result indicates the degree centrality differentials at the entire graph/network level. Much like the variance/standard deviation, which is a measure of individual score differences from the mean for the entire sample of cases, centralization measures the difference of actor centrality for the entire network/graph.

The formula for computing centralization is in the following.

$$C_G = \frac{\sum[C_D(N^*) - C_D(N_i)]}{(N-1)(N-2)}$$

C_G indicates the centralization measure for a given graph. $C_D(N^*)$ denotes the degree centrality for the node with the highest degree centrality. $C_D(N_i)$ indicates the degree centrality for all other nodes (nodes that do not have the highest degree centrality). The numerator $\sum[C_D(N^*) - C_D(N_i)]$ dictates that the node with the highest degree centrality minus the degree centrality of each of all other nodes in a given graph, and the differences are aggregated. The denominator $(N - 1)(N - 2)$ is the mathematical maximum value for the most unevenly distributed degree centrality among all nodes in a given graph.

To illustrate, I conjure up three artificial graphs (all are binary undirected for simplicity). Figure 3.12 shows the three networks with the graph on the left being the "star graph." The graph on the right is the wagon wheel, and the graph in the middle is somewhere in between of these two extreme configurations. Applying the centralization measure formula to the three graphs generates their centralization measures as 100%, 41.67%, and 0% respectively. In detail, applying the formula to the star graph, node E has the highest degree centrality as 4, and all other nodes have the same degree centrality as 1. The numerator would be $(4 - 1) + (4 - 1) + (4 - 1) + (4 - 1)$, which is 12. And the denominator is $(5 - 1(5 - 2) = 12$. Thus, the centralization is 100% ($\frac{12}{12} \times 100$). Notice that the star graph and the wagon wheel are directly contradictory to each other in their distributions of degree centrality of all nodes. While the wagon wheel has the most equal distribution, star graph has the

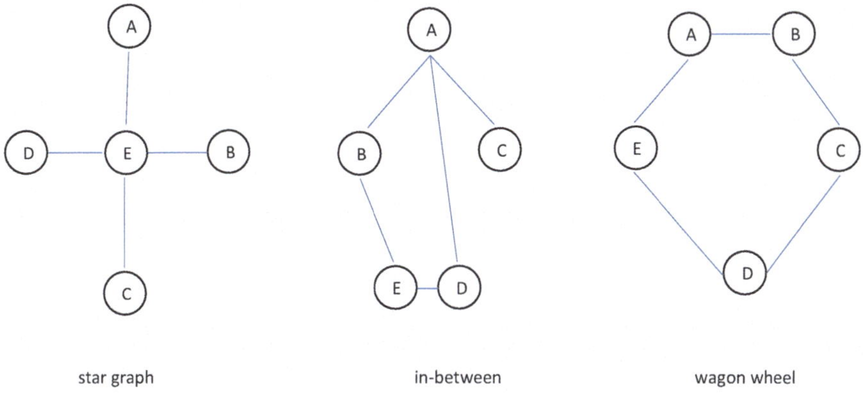

Fig. 3.12 Three centralizations in binary undirected graphs

most uneven distribution. Therefore, the closer a centralization is to 100%, the more unequal or uneven the distribution of degree centrality is for the entire graph. Conversely, the closer a centralization is to 0%, the more equal or egalitarian the distribution of degree centrality is. Imagine a workplace communication network where in scenario A, everybody is talking to everybody else, and in scenario B, everybody is talking to one person only. A's network centralization would be 0%, where B's network centralization would be 100%. Most real life situations are sitting somewhere in the middle (such as the graph in the middle of Fig. 3.12).

References

1. Burt RS (2004) Structural holes and good ideas. Am J Sociol 110(2):349–399
2. Burt RS (2007) Brokerage and closure: an introduction to social capital. Oxford University Press, Oxford
3. Burt RS (1992) Structural holes: the social structure of competition. Harvard University Press, Cambridge, MA
4. Crossley N, Bellotti E, Edwards G, Everett MG, Koskinen JH, Tranmer M (2015) Social network analysis for ego-nets. Sage, Thousand Oaks
5. Knoke D, Burt RS (1983) Prominence. In: Burt RS, Minor MJ (eds) Applied network analysis. Sage, Beverly Hills, pp 195–222
6. Marsden PV (2011) Survey methods for network data. In: Scott J, Carrington P (eds) The Sage handbook of social network analysis. Sage Publications, London, pp 370–388
7. Yang S (2022) Social network analysis. In: Frey BB (ed) The SAGE encyclopedia of research design. Sage, Thousand Oaks, pp 1546–1551

Chapter 4
Advanced Method in Social Network Analysis

Song Yang

Abstract In previous chapters, we discuss the social network analysis basics such as nodal degree centrality, measures of dyadic pairs, cliques and subgroups, as well as centralization and density measures at the whole network level. This chapter addresses more advanced topics such as mapping and clustering, bipartite network, and ERGM. It starts with the mapping technique such as MDS and hierarchical clustering to depict the structuration and configuration of social networks. Then it introduces bipartite network that involves heterogenous sets of nodes in social network analysis. It concludes with a discussion of Exponential Random Graph Modeling (ERGM), an unique model method that tries to explain the social network outcomes.

Keywords Multidimensional Scaling (MDS) · Hierarchical clustering · Bipartite network · Exponential Random Graph Modeling (ERGM)

In previous chapters, we discuss the social network analysis basics such as nodal degree centrality, measures of dyadic pairs, cliques and subgroups, as well as centralization and density measures at the whole network level. This chapter addresses more advanced topics such as mapping and clustering, bipartite network, and ERGM. It starts with the mapping technique such as MDS and hierarchical clustering to depict the structuration and configuration of social networks. Then it introduces bipartite network that involves heterogenous sets of nodes in social network analysis. It concludes with a discussion of Exponential Random Graph Modeling (ERGM), an unique model method that tries to explain the social network outcomes.

One of the unique advantages of engaging social network analysis lies in its ability to depict the complex social network structures in vividly visualization pictures. Multidimensional scaling (MDS) is one of those techniques that widely used in social network analysis to reveal social network with visualization mappings. It is often used in connection with hierarchical clustering to depict the overall

S. Yang (✉)
Department of Sociology and Criminology, University of Arkansas, Fayetteville, AR, USA
e-mail: yangwang@uark.edu

S. Yang (ed.), *Social Network Analysis in Action*, Lecture Notes in Social Networks,
https://doi.org/10.1007/978-3-031-66661-2_4

structuration as well as local cliques and strongly connected groups. Sections below will discuss both methods of MDS and dendrogram of hierarchical clustering.

4.1 Multidimensional Scaling (MDS) and Hierarchical Clustering

Multidimensional scaling (MDS) is a visualization tool to map the distance between nodes in a network [6, 12, 13]. The input data structure of multidimensional scaling is commonly a squared matrix or adjacency matrix, with entries indicating the distance/proximity between their corresponding nodes. The output of MDS is a spatial presentation, consisting of a geometric configuration of points. Normally the larger the similarities (or the smaller the dissimilarities) between nodes, the smaller the distance of them in the configuration map.

Assuming that we have five nodes, and their coordinates are displayed in a 2-dimensional plane, we would first produce the symmetric valued matrix using the following formula:

For a given pair of nodes between i, and j, their Euclidean distance on a 2-dimensional plane is

$$d_{i,j} = \sqrt{(x_i - x_j)^2 + (y_i - y_j)^2}$$

The above formula can be used repeatedly for all the pairs that exist between the five nodes, which totals 10 dyadic pairs ($C_5^2 = \frac{5!}{2! \times 3!} = 10$). Thus, we would have a matrix with the entries indicating the distances between the nodes

$$\begin{bmatrix} d_{1,1} & d_{1,2} & d_{1,3} & d_{1,4} & d_{1,5} \\ d_{2,1} & d_{2,2} & d_{2,3} & d_{2,4} & d_{2,5} \\ d_{3,1} & d_{3,2} & d_{3,3} & d_{3,4} & d_{3,5} \\ d_{4,1} & d_{4,2} & d_{4,3} & d_{4,4} & d_{4,5} \\ d_{5,1} & d_{5,2} & d_{5,3} & d_{5,4} & d_{5,5} \end{bmatrix}$$

The above matrix is a valued symmetric matrix with the diagonal values equal 0:

$$d_{i,j} = d_{j,i}$$

$$d_{i,i} = 0$$

When the above matrix is input to SNA software (such as UCINET) for MDS graphing, it will generate a map to show the proximities between the nodes. In addition to the mapping, the MDS function also produces ***a stress indicator*** [6]. The stress indicator computes the discrepancies across all pairs between the observed matrix and the computed matrix,

$$stress = \sqrt{\frac{\sum\sum \left(f\left(x_{i,j}\right) - d_{i,j}\right)^2}{scale}}$$

$f(x_{i,\ j})$ is the computed distance between a given pair, whereas $d_{i,\ j}$ is the observed Euclidean distance between the pair. The denominator of scale is a constraining factor to ensure the stress indicator to be between 0 and 1. As the best-case scenario, the reproduced map perfectly replicates the original distance $f(x_{i,\ j}) = d_{i,\ j}$ for all pairs; the stress indicator would be 0. Conversely, when the reproduced distances between the pairs have sizable differences from their original Euclidean distances for all pairs, the stress indicator reaches 1. For this reason, the stress indicator is dubbed a "badness of fit" indicator, with the larger the value, the worse the replicated map in representing the original distances. Scholars [6] recommend the following guidelines in interpreting the stress indicator:

$$0 \le stress\ indicator \le 0.1 \quad \text{Excellent fit}$$

$$0.1 < stress\ indicator \le 0.2 \quad \text{Adequate fit}$$

$$0.2 < stress\ indicator \le 1 \quad \text{Poor fit}$$

To illustrate, I conjure up an artificial social network of co-authorship among 7 departmental members. Table 4.1 shows the adjacency (squared) matrix depicting the patterns of co-authorship, with 1 indicating the presence and 0 indicating the absence for a given pair of nodes. In addition, the data also have faculty members' biostatistics such as gender and race. Among the total 7 faculty members, five are men (indicated by yellow color) and two are women (indicated by green color). Along racial dimension, five are White (indicated by square) and two are people of color (indicated by circle). The information of faculty's biostatistics (gender and race) will be used as exogenous variables in later discussions of Exponential Random Graph Modeling (ERGM).

Figure 4.1 shows the MDS processing with Stata MDS solution. Noted that the co-authorship network is very well connected, many pairs have direct ties, and the ones not directly connected are connected via no more than two steps. The enclosed circles identify small groups that are strongly connected with each other. One of those is a clique, involving three members TS-HC-MN. The other is a pair,

Table 4.1 A hypothetical network of co-authorship between 7 faculty members

	HC	TS	SM	SY	JB	MN	FK
HC	0	1	0	0	0	1	1
TS	1	0	0	0	1	1	0
SM	0	0	0	1	1	1	0
SY	0	0	1	0	0	1	0
JB	0	1	1	0	0	0	0
MN	1	1	1	1	0	0	1
FK	1	0	0	0	0	1	0

Fig. 4.1 STATA SNA/MDS solution of the aforementioned co-authorship network (Stress = 0.2). Note: Colors represent gender (green-women, yellow-men) and shapes represent race (square-white, circle-people of color)

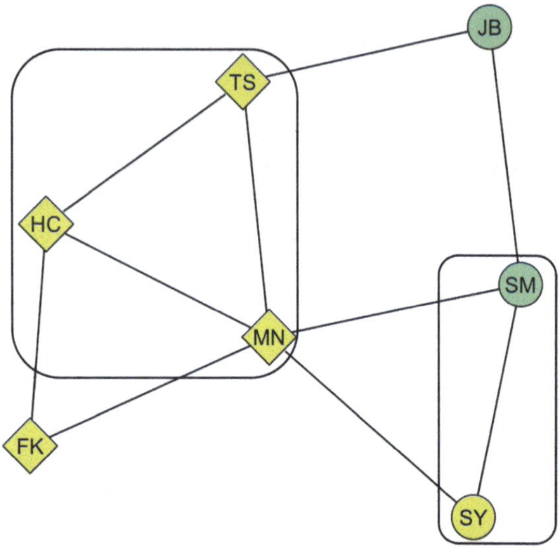

connecting SM with SY. The enclosed circles are dendograms that are based on the hierarchical clustering analysis of the adjacency matrix, which is discussed in the following paragraphs. The MDS of the co-authorship network has overall stress factor of 0.2, suggesting adequate accuracy in the reproduced graph in representing original one.

4.1.1 Dendogram: A Method of Agglomerative Hierarchical Clustering

A dendogram is one type of visual presentation of applying the agglomerative hierarchical clustering to a social network dataset. Hierarchical clustering is an agglomerative method in which we start with individual nodes of a network and join them to form groups. The basic idea behind the hierarchical clustering is to define a measure of similarity or connection strength between nodes, and then join the closest or most similar nodes to form groups. One issue of using hierarchical clustering is how to define similarity—in practice, one common choice is using correlation coefficients, and another one is Euclidean distance. Once the similarity measure is chosen, we need to compute the measures for all pairs of nodes in a network and start the clustering processes.

MDS and dendograms are often used in combination to show the overall network structure and the clustering or subgrouping in the network. In MDS, there is no differentiation of groups or cliques; dendograms thus complement MDS by identifying or isolating different cliques. One approach of such combined use of both is to draw a line at the dendogram that shows the snapshot of the clustering between the

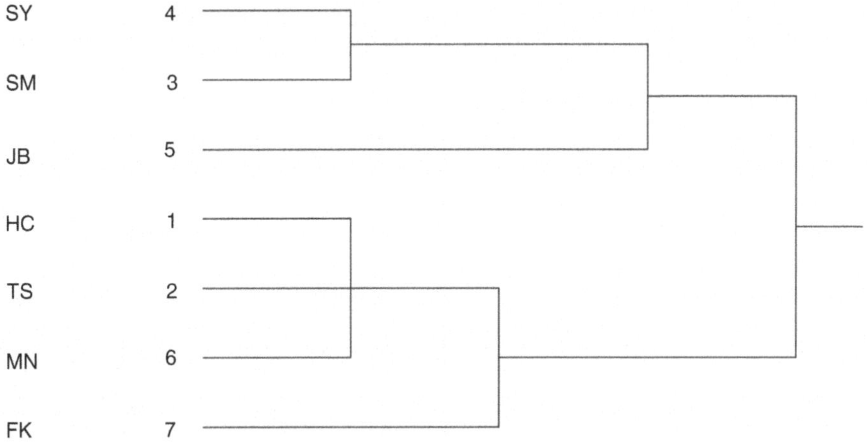

Fig. 4.2 Dendrogram of co-authorship network in Table 4.1 from hierarchical clustering

nodes, reflecting such clustering patterns in the MDS. The benefit of such an approach is that it provides a clear snapshot of the grouping in the network, whereas the drawback is that it is only a static picture, ignoring the dynamic processing of the agglomerative clustering. The second approach requires researchers to use some animation or simulation feature of some presentation software (such as customized animation in PPT) to illustrate the agglomerative hierarchical clustering. Such an approach can show the dynamic process of clustering. Figure 4.2 shows the hierarchical clustering of aforementioned co-authorship network. Looking from left to right in the graph, the algorithm first merges two clusters "SY-SM" and "HC-TS-MN." This is what the enclosed circles in Fig. 4.1 are based on. Then the threshold to conjoin cluster is lower, FK is joined to the "HC-TS-MN" cluster. Keep going right as the threshold is further decreasing, JB is joined with the "SY-SM" cluster. At the last step, everybody in the network is joined together.

4.2 Bipartite/Two-Mode Networks

Ever heard the new definition of CEOs? CEOs are those who discuss business in golf courses and who discuss golf in their offices. It is actually truer than it sounds. That informal network among CEOs is extremely important to get business done, either for the organization they represent or for their own personal advances. Anecdotal examples abound, recent sport news of men's basketball coach carousel vividly illustrates the important of social networking for employers and employees to find each other, even at the very top level. While Razorback men's basketball lost its competitive coach Eric Musselman and all of its players, it successfully attracts the legendary blueblood coach John Calipari from Kentucky within 24 hours. How such a millions of dollars contract and high level personnel move was cemented in such short and feverish fashion? This all thanks to the long time Razorback booster billionaire John Tyson, founder of Tyson Food, who also has friendship with John Calipari [8].

Bipartite networks/graph or two-mode networks include nodes that represent two distinctive sets of units. In the above anecdotal example, the two modes are persons (athletic directors, boosters, and coaches), and country clubs where personal friendship is cemented. In academic pursuits, Galaskiewicz's [3] used bipartite network to study 26 CEOs and their 15 social clubs. Freeman, Romney and Freeman's [2] applied bipartite network to studies of nine faculty, student, and research colloquiums at a university campus, and the observation of regular beach-goers over 31 days to reveal the subgroups. The classic study of bipartite networks goes back to 1941 when Davis and associates collected data on 18 southern women participating in 14 social events and discovered that the women clearly split into two camps [1]. Bipartite network studies are particularly useful for studying situations in which respondents cannot be interviewed or would be unlikely to report truthfully. In some cultures, respondents may feel compelled to report, e.g. friendship with everybody else in the same class or the same workplaces. In such a situation, observing actual behavior and co-attendance of events can help reveal the social structure of a group of classmates or coworkers.

Bipartite network can be converted to two different one-mode networks through matrix multiplication. To illustrate, I conjure up an artificial example of CEOs and Golfing events. Figure 4.3 shows that five CEOs (A, B, C, D, and E) attend four golf events (1, 2, 3, and 4). Table 4.2 demonstrates the bipartite network in Fig. 4.3 in the matrix format. In the upper portion of the matrix where CEOs are linked to Golfs, participation of the CEOs (in rows) at Golf events (in columns) is indicated by "1", nonparticipation is indicated by "0". In the lower portion of the matrix where Golf is related to CEOs, the entries of the matrix simply indicate the golf events (in the rows) attract CEOs (in the columns). Noted that both the CEOs by CEOs and Golf by Golf

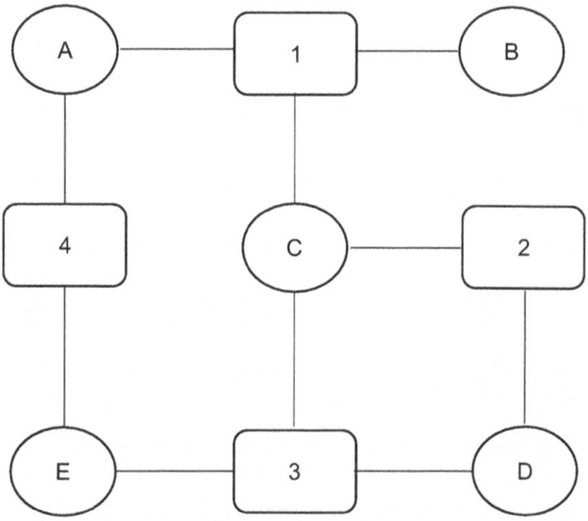

Fig. 4.3 Bipartite network of CEOs and Golfing

Table 4.2 The Matrix of bipartite network of CEOs and Golfing

	A	B	C	D	E	1	2	3	4
A						1	0	0	1
B						1	0	0	0
C						1	1	1	0
D						0	1	1	0
E						0	0	1	1
1	1	1	1	0	0				
2	0	0	1	1	0				
3	0	0	1	1	1				
4	1	0	0	0	1				

matrices in the table are empty. They will be filled out once the conversion of bipartite network to one-mode network is completed through metrices multiplications.

A bipartite matrix can be turned into one-mode matrix to reveal the network structure between social actors or between social events. Let's assume the upper matrix in Table 4.2 to be A, and A' is the transpose of the matrix A, which is shown in lower portion of Table 4.2. Basically, the upper matrix of A is the CEOs by Golf, where the lower matrix of A' is Golfs by CEOs. Multiplying A with A' produces an adjacency matrix of the social network between the five CEOs through their co-participation of Golf events.

$$AA' = \begin{pmatrix} 1\,0\,0\,1 \\ 1\,0\,0\,0 \\ 1\,1\,1\,0 \\ 0\,1\,1\,0 \\ 0\,0\,1\,1 \end{pmatrix} \begin{pmatrix} 1\,1\,1\,0\,0 \\ 0\,0\,1\,1\,0 \\ 0\,0\,1\,1\,1 \\ 1\,0\,0\,0\,1 \end{pmatrix} = \begin{pmatrix} 2\,1\,1\,0\,1 \\ 1\,1\,1\,0\,0 \\ 1\,1\,3\,2\,1 \\ 0\,0\,2\,2\,1 \\ 1\,0\,1\,1\,2 \end{pmatrix}$$

The resulting matrix depicts the CEOs-by-CEOs connections, which is shown in the left table in Table 4.3. Compared with the conventional squared (adjacency) matrix, such one-mode matrix is much more informative. First, the diagonal values in the one-mode matrix have substantive meanings. In this case, they mean the total number of golf events attended by the CEOs in the rows. For example, CEOs A, B, C, D, and E attend 2, 1, 3, 2, and 2 golf events respectively. Second, this one mode matrix is an undirected symmetric matrix in which $X_{i,\,j} = X_{j,\,i}$. The off-diagonal entries indicate the total number of golf events attended by CEOs *i* and *j*. For example, CEOs A and B co-attended one golf event, whereas A and D co-attended no golf events. The higher the number, the more the golf events co-attended by pairs of CEOs. Such matrix can provide a proxy to reveal the social connections between pairs of CEOs.

Feeding such one-mode CEO by CEO matrix into MDS solution produces a graph showing social networking among the five CEOs based on their co-participation in golf events. Figure 4.4 shows that the graph reaches stress factor at 0.04, indicating excellent fit between reproduced graph and the original network. Dendrogram based on hierarchical clustering produces only one cluster (CEOs A

Table 4.3 The matrices of CEOs and Golfs

CEO by CEO Matrix					
	A	B	C	D	E
A	2	1	1	0	1
B	1	1	1	0	0
C	1	1	3	2	1
D	0	0	2	2	1
E	1	0	1	1	2

Golf by Golf Matrix				
	1	2	3	4
1	3	1	1	0
2	1	2	2	0
3	1	2	3	1
4	1	0	1	2

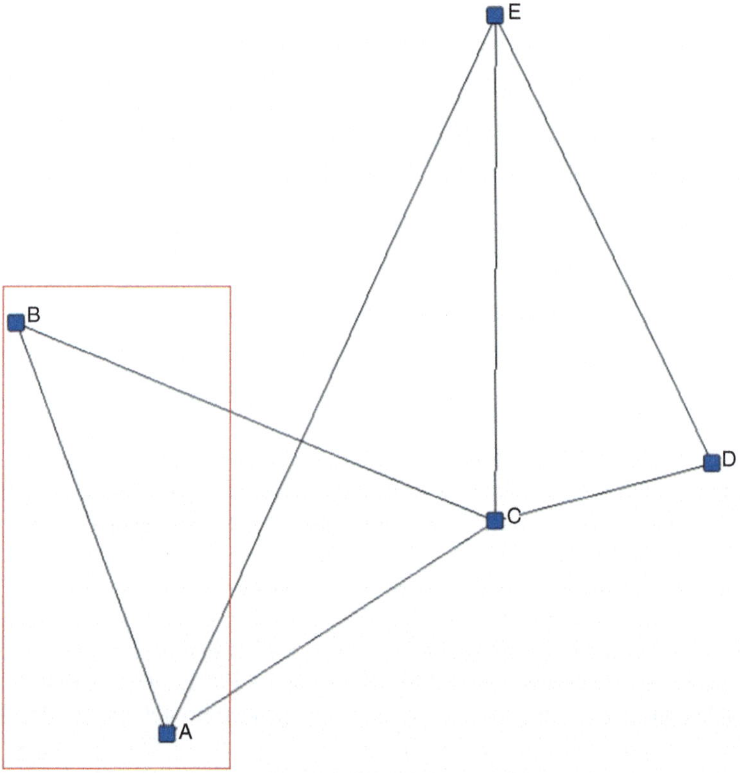

Fig. 4.4 MDS of the one-mode network between CEOs (Stress factor = 0.04)

and B) before everybody else joins the group. CEO C turns out to be the most critical player in this small network as it connects with all other CEOs, having the highest centrality in the network.

Conversely, if one multiplies A′ with A in Table 4.2, it will produce a 4 by 4 "golf events by golf events" matrix, which entails a different interpretation than it is in the previous matrices' multiplication. The following matrices multiplication shows the process.

$$A'A = \begin{pmatrix} 1\,1\,1\,0\,0 \\ 0\,0\,1\,1\,0 \\ 0\,0\,1\,1\,1 \\ 1\,0\,0\,0\,1 \end{pmatrix} \begin{pmatrix} 1\,0\,0\,1 \\ 1\,0\,0\,0 \\ 1\,1\,1\,0 \\ 0\,1\,1\,0 \\ 0\,0\,1\,1 \end{pmatrix} = \begin{pmatrix} 3\,1\,1\,0 \\ 1\,2\,2\,0 \\ 1\,2\,3\,1 \\ 1\,0\,1\,2 \end{pmatrix}$$

The right matrix in Table 4.3 shows the golf event by event matrix. Similar to the one mode network of CEOs by CEOs, this one mode event-by-event matrix has diagonal values with substantive meanings, which indicate the total number of CEOs who attended that particular golf event. Golf events 1, 2, 3, and 4 attract 3, 2, 3, and 2 CEOs respectively. Those off-diagonal values indicate the number of CEOs shared by the two golf events. For example, golf events 1 and 4 share no CEO, whereas golf events 2 and 3 share 2 CEOs.

Some interesting statistics can be readily applied to the two computed matrices to yield additional information about the structures. Taking the mean of the diagonal values of the CEO matrix would produce the average number of golf events attended by each CEO ($\overline{X^A} = \frac{\sum_{i=1}^{g} x_{i,i}}{g}$). Similarly, the mean of the diagonal values of the golf event matrix produces the average number of CEOs drawn to each event ($\overline{X^E} = \frac{\sum_{i=1}^{E} x_{i,i}}{E}$). The two numbers for our example above are $10/5 = 2.0$ and $10/4 = 2.5$, respectively, suggesting that on the average, each CEO attends 2 golf events, and each golf event draws 2.50 CEOs.

After converting the bipartite network into one-mode network, those one-mode networks (shown in Table 4.3) can be subject to analyses with measures discussed before, such as density, centrality, centralization, and cliques. One issue worth mentioning is that the one-mode matrix after converting from the two-mode/bipartite matrix is always undirected or symmetric, as the event sharing, or actor sharing does not distinguish between who is the sender and who the receiver. In addition, the values in the matrix can exceed binary range, with higher numbers indicating more shared events by pairs of actors, or shared actors by pairs of events. Some analytics would binarize the valued entries to simplify the social network analysis, such practices would sacrifice the rich data in favor of easier analysis and straightforward interpretations.

4.2.1 Exponential Random Graph Modeling (ERGM)

For decades in the development of social network analysis methods, researchers have been trying to come up with a satisfactory framework for explanatory purpose. Noted that most methods we discussed so far are descriptive, describing the nodes, cliques, dyadic pairs, and the entire network with data reduction techniques. Explanatory approach, which attempts to explain the underlying reasons for the formation of ties and the network outcomes has been limited, and preliminary until early 2010s when computer algorithm was developed and implemented to offer readily available software to analyze social network data [7]. The following sections will discuss this ERGM methods in details, starting with the unique features of social network data that challenge the conventional explanatory methods. Due to the complexity of this methodology, I draft this section with the most straightforward way in layman's terms. Audiences craving for more deep understanding of the subject can consult with writings such as [5, 7].

Before I discuss the ERGM in detail, I need to address the issues with using conventional statistical methods on explaining social network outcomes. After all, many studies before the availability of ERGM with R packet or Siena 4.0 [9] were using conventional software to analyze social network data. Such practices have three major problems that make them unsatisfactory solutions. The first lies in the nature of social network data of being interdependent of each other. Such innate characteristics of social network data runs counter to the basic assumption of conventional statistical packet that cases in a dataset must be independent of each other. Think about the famous social survey data of General Social Survey or GSS [4] (https://gss.norc.org/), cases are randomly selected individuals from US population that are independent of each other: the chance of any one case to know others and influence others is little to none. However, complete social network data often come from a small group with well defined boundary. In complete social networks, not only cases are known to each other, but also any individual's actions would have rippling effects on the rest of the network, hence affecting the overall structure of the network.

The second problem is closely related to the first one of the data being interdependent. That is: in social network data, some structural characteristics are more likely to occur than their total random chances. For example, reciprocity or mutuality is highly present in many dyadic pairs, or in other words: the chance of those mutuality to occur is much higher than they are to occur total randomly. Other structural features similarly to the reciprocity are two stars, and transitivity (triangle). Figure 4.5 shows those local network configurations/processes that are prevalent in social network data [7]. Those local network configurations need to be accounted for before the net effects of other variables (e.g. node or social actor characteristics) on the formation of network ties can be estimated. Conventional statistical packets are unable to assess and control for the effects of those local network configurations.

The third problem with running conventional explanatory modeling on social network data is related to the inferential statistics. Complete social network data

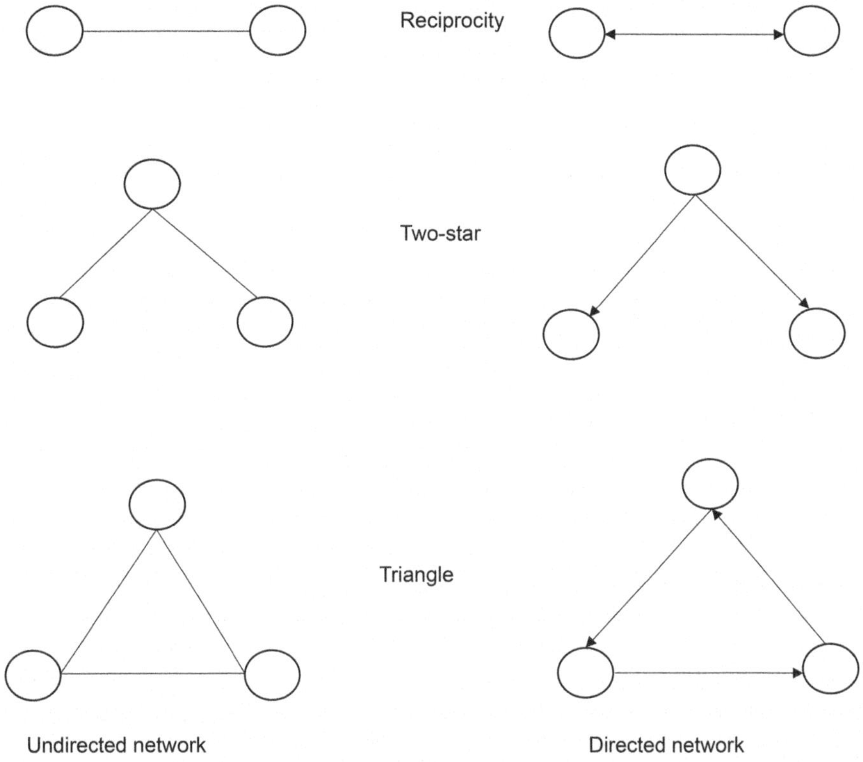

Fig. 4.5 Common local network configurations in ERGM controls

rarely comes from random sampling of open population (think of GSS for example). Instead, they are either the entire subpopulation with a well-defined boundary (students in a class, CEOs in a country club, employees with an organization etc.), or their targeted population is unknown. In those cases, conventional inferential statistics such as chi-square (χ^2) test, Z test, student t test, or F ratio test, are not appropriate to infer the representativeness of social network data. Some innovative approach must be invented to assess the overall network modeling as well as individual parameters in terms of their representativeness of the population.

Exponential random graph model (ERGM) attempts to explain how and why social network ties arise. The main goal of ERGMs is to understand a given observed network, that is the empirical network measured by researchers, and to obtain insight into the underlying process that creates and sustains its ties. ERGMs represent a network (graph G) through various network statistics or characteristics that make specific formations or ties more likely, such as mutuality, or transitivity. An ERGM assigns probability to graphs according to the following statistics ([7]: 9),

$$P_\theta(G) = Ce^{\theta_1 Z_1(G)+\theta_2 Z_2(G)+\theta_3 Z_3(G)+\ldots\theta_p Z_p(G)}$$

Such Formula stipulates that the probability of observing the specific network G is given by the sum of network statistics (the Zs in the expression, which are similar to the variables in conventional multiple regression), weighted by the parameters (the θs) in the exponential function, in which c is a normalizing constant. Those network statistics exemplify a set of network configurations, which are small local subgraphs in the network. The parameter attached to each configuration informs us of the importance of the configuration.

Such model should address all three concerns raised above. First, it allows researchers to explicitly control for the effects of data interdependency by including in the model the common tendencies such as mutuality, two-star, and transitivity. Second, the model is able to assess the statistical significance of parameters through nonparametric method of simulations, rather than relying on parametric assumption of normality distribution of sampling. The clear advantage of this ERG model lies in its ability to assess multiple independent variables (the Zs), their respective effects (the θs) and their significant levels in their role in producing the observed social network outcomes. For those who are familiar with non-network multiple regressions, such as logistic regression, interpretations of ERGM parameters are very similar to those in the logistic regression (assuming the dependent matrix is the binary), although their underlying principles and machine algorithms are very different. Such feature helps greatly to propagate ERGM among the majority of researchers who are intimidated by ERGM but are familiar with the multiple regression (logistic regression). As long as researchers know the usage of specialty software package of ERGM, such as R Statnet [10], Siena [11], or pNet [7], they can easily apply ERGM to their substantive research areas.

Statistical modeling with Exponential Random Graphs investigates the observed network, and tests if particular network structures are significantly more common than would be expected by chance. A nonparametric statistical technique to address this question is to simulate a large number of random graphs and compare the count of those network structures in simulated random graphs with that in the actual observed network. The distribution of the counts in the simulated random networks is called sampling distribution, analogous to the conventional sampling distribution in parametric statistics. The count in the observed network structure is then compared with that of the sampling distribution, which in turn produces statistics, such as confident intervals and significant levels of the observed network structures. In this vein, Chap. 2 in Lusher et al. [7] provided an artificial example of the statical simulation that readers can consult.

Statistical modeling tries to find a distribution of graphs in which the observed graph is the center of the distribution. Then, it will fit the model by estimating parameters. For a finite set of nodes, a sampling distribution of ties between the nodes can be very large. In particular, for a network with size N, the total possible arrangement for possible network states is $2^{(N)(N-1)}$. Let's start with the simplest network with only 2 nodes, which would have $2^{2(1)} = 4$ network states. A network with 3 nodes would have 64 different states ($2^{3(2)} = 64$). The number becomes extremely large with a very slight increase in the size of the network—a network with merely 5 nodes would have $2^{20} = 1{,}048{,}576$ states. Simulating all possible

Table 4.4 ERGM of co-authorship network among 7 faculty members in Table 4.1

	Model 1	Model 2	Model 3
Reciprocity 	2.394 (2.1)	—	−2.665 (3.605)
Two-star 	−0.454 (.539)	—	.816 (.999)
Triangle 	−0.322 (1.083)	—	−3.537 (2.42)
Race	—	1.537 (1.092)	3.869* (2.129)
Gender	—	−.358 (.911)	1.509 (1.421)
Model loglikelihood	−13.51	−13.00	−9.256
Model AIC	33.03	30.01	28.51

*P < .05

states of a large network thus requires powerful computers and fast algorithms. For the distribution of the graphs, ERGM assigns a probability to each of the realizations of random graphs, so that the observed graph is in the center of this distribution of random graphs. The particular algorithm that was used in R Statnet in ERGM is Markov chain Monte Carlo (MCMC), which produces statistics of individual parameters in the model, as well as overall model fitness. To illustrate the ERGM, I use R Statnet ERGM module to estimate the effects on the co-authorship network, displayed in Table 4.1 and Fig. 4.1.

The co-authorship network of 7 faculty members has an overall network density of 47.6 percent ($\frac{10}{C_7^2} = \frac{10}{21} = 47.6\%$), indicating good connections of co-authorship between the faculty members. In addition, as the Fig. 4.1 shows, it also has information about the faculty members' gender and race. With those, I develop three models and run ERGM on each of the three models. Model 1 is the baseline, examining the impacts on the formation of co-authorship of three local configurations (reciprocity, 2-star and triangle). Model 2 is the actor attribute model, examining the effects on the co-authorship of gender and race of faculty members. Model 3 is the overall model with both local network configurations and faculty's personal characteristics (gender and race) included. Table 4.4 summarizes the key findings, with maximum pseudo-likelihood estimates (MPLE) for each variable and their corresponding standard errors and p-values. Due to the small sample size, none of the models exhibits significance at the level of 0.05. However, in Model 3, with the local configurations being controlled, race exhibits certain level of homophily (p-value = 0.069, which is one-tail significance), implying people in the same race group (white vs others) tend to have more scientific collaborations (odds ratio $exp^{(3.869)} = 47.9$) then do people across racial lines.

In addition to the individual parameters, the R Statnet ERGM also produces estimates of model statistics such as model likelihood and model Akaike Information Criterion (AIC). These two measures indicate the level of discrepancy between the simulated graph and the observed one. In analogy, they are like the model errors of residue that sums up the discrepancy between the estimated values and the observed values in OLS regression. Basically, the lower the value in loglikelihood and AIC, the better the model fitness to the original graph. It appears that the model 3 with both the network endogenous configurations and actor attributes is the best fitting model of the three.

References

1. Davis A, Gardner BB, Gardner M (1941) Deep south: a social anthropological study of caste and class. University of Chicago Press, Chicago
2. Freeman LC, Romney K, Freeman S (1987) Cognitive structure and informant accuracy. Am Anthropol 89(2):310–325
3. Galaskiewicz J (1985) Social organization of an urban grants economy: a study of business philanthropy and nonprofit organizations. Academic, Orlando
4. General Social Survey or GSS. https://gss.norc.org/, site visited in May 2024
5. Harris JK (2014) An introduction to exponential random graph modeling. Sage, Thousand Oaks
6. Kruskal JB, Wish M (1978) Multidimensional scaling. Sage Publications, Beverly Hills
7. Lusher D, Koskinen J, Robins G (2012) Exponential random graph models for social networks: theory, methods, and applications. Cambridge University Press, Cambridge
8. New York Post. https://nypost.com/2024/04/08/sports/a-chicken-empire-and-bad-feelings-behind-john-calipari-leaving-kentucky-for-arkansas/, site visited on April 8, 2024
9. Ripley R, Snijders TAB, Boda Z, Vörös A, Preciado P (2017) Manual for SIENA version 4.0. University of Oxford, Department of Statistics, Oxford
10. Statnet Development Team. https://statnet.org/about/, site visited in May 2024
11. Snijders TAB. https://www.stats.ox.ac.uk/~snijders/siena/, site visited in May 2024
12. Wasserman S, Faust K (1994) Social network analysis: methods and applications. Cambridge University Press, Cambridge/New York
13. Yang S, Keller FB, Zheng L (2016) Social network analysis: methods and examples. Sage, Thousand Oaks

Chapter 5
A Whole Network Study: Gender and Racial Homophily in Classroom Discussion Network

Song Yang, Regan Hodges, Joseph W. Sirrianni, and Qingyang Zhang

Abstract The previous chapters describe various aspects of conducting social network analysis, from theoretical motivations, to research design, to data collections, and to data analytics. In this and the next chapter, I will discuss the research outcomes from data acquired, processed, and analyzed from the Intelligent Cyber Argumentation System (ICAS), which was discussed briefly in Chap. 2. This chapter uses whole network data scraped from ICAS on a controversial social issue: guns on campus to investigate the issue of racial and gender homophily on a controversional social issue: guns on campus. The flow of this chapter may look a bit different from the previous four chapters because this chapter follows the format of social science manuscript, rather than book monograph. It starts with an introduction, moving on to literatures, research questions, hypotheses, data and variables, findings, and finishes with conclusive remarks.

Keywords Whole network analysis · Gender/racial homophily · Controversial social issues · Guns on campus

The previous chapters describe various aspects of conducting social network analysis, from theoretical motivations, to research design, to data collections, and to data analytics. In this and the next chapter, I will discuss the research outcomes from data acquired, processed, and analyzed from the Intelligent Cyber Argumentation System (ICAS), which was discussed briefly in Chap. 2. This chapter uses whole network

S. Yang (✉)
Department of Sociology and Criminology, University of Arkansas, Fayetteville, AR, USA
e-mail: yangwang@uark.edu

R. Hodges
Simmons Foods, Rogers, AR, USA

J. W. Sirrianni
Department of Computer Science and Computer Engineering, University of Arkansas, Fayetteville, AR, USA

Q. Zhang
Department of Mathematical Sciences, University of Arkansas, Fayetteville, Arkansas, USA

S. Yang (ed.), *Social Network Analysis in Action*, Lecture Notes in Social Networks,
https://doi.org/10.1007/978-3-031-66661-2_5

data scraped from ICAS on a controversial social issue: guns on campus to investigate the issue of racial and gender homophily on a controversional social issue: guns on campus. The flow of this chapter may look a bit different from the previous four chapters because this chapter follows the format of social science manuscript, rather than book monograph. It starts with an introduction, moving on to literatures, research questions, hypotheses, data and variables, findings, and finishes with conclusive remarks.

In particular, research findings in this chapter show that two types of network emerged: the positive edge network, in which participants are mutually supportive of each other's viewpoints, and the negative edge network, in which participants have disagreement and disputes with each other. In aidditon, the study reveals significant gender gaps in negative edge network occur that cross-gender dyads and female-female dyads are less likely to engage in debate in negative edge network than are male-male dyads. Compared with white and white dyads, interracial dyads between white and Hispanics are less likely to engage in discussion in overall network. Interracial dyads between white and Asians are less likely to engage in debate in negative edge network.

5.1 Introduction

It has long been established that homophily is a powerful social force shaping one's social ties and network environment [20]. Although homophily occurs in many dimensions, such as age, education, occupation, income levels, religion, and family backgrounds, the most important dimensions of homophily are the basic demographic features of gender and race. This is especially true in a multiracial and multicultural society such as the U.S., where people with diverse racial/ethnicity and cultural background try to live and work harmoniously through inter-racial assimilation or accommodations. However, when it comes to inter-personal associations, such as marriage, friendship, advice, support, information transfer, exchange, comembership, religious service, or even group exercises, racial homophily unleashes its powerful force to bind the same race together while at the same time, deter and dissolve cross-racial interactions.

Gender homophily, though it occurs less frequently than racial homophily, also emerges from time to time. For instance, gender homophily occurs in workplaces, where women unite together in their collective effort to break the glass ceiling [5]. Gender homophily is a common occurrence in male-dominated occupations from manual labor jobs, such as coal mines, to high prestigious occupations, such as physicians. In those occupational settings, gender-based associations serve as a gatekeeping mechanism or to establish a threshold to keep "the others" out. It is also worth noting that although gender homophily sometimes operates solo, it often bundles with racial homophily, creating powerful homogenous gender/racial coalitions [31].

In this research, we investigate the presence of racial and gender homophily in an online classroom discussion on the issue of allowing guns on a college campus. This guns on campus issue, whether or not to allow concealed carry of personal weapons on university campus, generated heated debates among students in a classroom of a large public land-grant research university in South East Conference (SEC). Fierce disagreements broke out between the conservative pro-gun group and the liberal group of anti-guns. A unique aspect of our research is that, since the discussions took place online, we were able to make the participants anonymous, such that the participants did not know each other's identities during the discussion. The resulting discussion network is purely a candid shot of how the participants engage in discussions without any knowledge of their discussion partners' demographic features.

Another distinguishing feature of our research is that we are able to separate the discussion network into two different types of networks, a positive edge network and a negative edge network, based on the users' interactions in the online discussion. The positive edge network only contains discussion interactions in which the participants are agreeing with each other, while the negative edge network only contains discussion interactions in which the participants are disagreeing with each other. It is highly critical to make this distinction between the interactions because the gender and racial homophily in discussion network of the controversial social issue unfolds very differently between the two types of networks, which we will show later. For the rest of this paper, we will be discussing the following topics: relevant literatures on gender and racial homophily, our research questions, research hypotheses, data collection process, variable constructions, methods used, and major findings. We conclude this paper with a summary of highlights, contributions to the existing literatures, and future research that can extend along this line of work.

5.2 Literatures on Gender and Racial Homophily

Homophily simply means that when engaging in social interaction, individuals tend to interact with others who are similar in characteristics. Homophily centers on the principle that interactions between similar people are more frequent than between individuals who are dissimilar. This phenomenon has become a focus of studies on social networks and can be traced back to classical Western writings (e.g. [1]). Systematic observations of network ties and group formation by social scientists first noted that children formed friendships at school when they shared similar demographics [2, 11, 30]. Most notably, Lazarsfeld and Merton [16] utilized Simmel's [25] classic theory to term "homophily" and quote that, "birds of a feather flock together," which is still used to describe the phenomenon.

In early studies of homophily, researchers focused on social ties between members of small groups, with most research centered on children. These studies provided evidence for the role of demographic characteristics, such as race, age, gender, and education [2, 19] in group formations. As issues of race became more prominent during the mid-century, the focus shifted toward association patterns based on race and ethnicity. By the late 1980s, advancements in technology allowed for researchers to sample larger groups of schools [6, 24] and communities [8, 15, 29], making findings more generalizable to the general population. This advancement in methodology eventually led to researchers being able to measure homophily across multiple characteristics.

5.2.1 Factors Conducive to Homophily

The most prominent component to research on homophily is race and ethnicity. Studies have found strong evidence of racial homophily in a variety of relationships, including friendships [24], co-workers [12, 17], and marriage [13]. While most evidence is in relation to larger groups, racial/ethnic homophily can also be observed in smaller groups, like the workplace [12] and in classrooms [10].

Much like the patterns of racial and ethnic ties, gender homophily appears to be prevalent in various settings. For example, gender homophily appears in education and in professional workplaces [23, 33]. In addition to homophily, some research studies suggest that gender also affects the process of forming social networks. For example, Eder and Hallinan [7] found that girls are more likely to resolve network tension by removing a social tie, while boys are more likely to add problematic social ties to their network. This difference in processes result in smaller networks for girls and larger networks for boys during adolescence but can be abated as individuals move into adulthood.

One of the most intriguing studies documenting gender differences in using network ties in career development appears in Burt's research of structural holes [3, 4]. In his classical thesis, Burt [3] stated that individuals connecting with others who are disconnected from each other benefit from such network configurations, or structural holes in his terminology. Figure 5.1 shows two network configurations: the graph on the left produces few structural holes but great level of social closure to the ego, while network on the right affords many structural holes to the ego in the graph. According to Burt, structural holes generate two benefits to the focal network

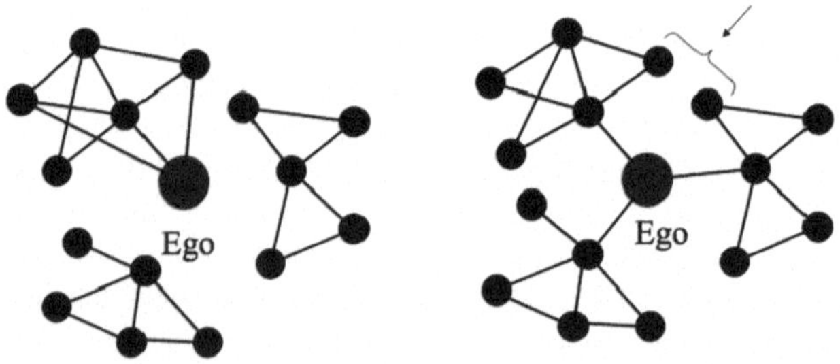

Fig. 5.1 Structural hole and social closure

actor: information benefits and control benefits. The information benefits refer to that network actors bridging many structural holes receive non-redundant, additive, and novel information and intelligence. The control benefits mean that network actors with many structural holes can also play their contacts off each other. For example, In a business context, network actors with many structural holes can exploit these by playing their contacts off each other to achieve better business deals. Hence, the ego (the focal network actor) on the right graph will receive those benefits resulting from the structural holes in its network, whereas the ego on the left will not receive those benefits due to its lack of structural holes.

The most fascinating part of Burt's research is that he later discovered that women do not benefit from the structural holes, those alleged benefits of structural holes only accrue to men [4]. In particular, in terms of improving the chances of receiving large pay raise and early promotion, men benefit from structural holes while women benefit from social closure. Simply put, the graph on the left in Fig. 5.1 produces benefits to women, while the graph on the right produces benefits to men. Why is this? Burt asserted two reasons: (1) the cultural reason that women tend to thrive in supportive groups that are associated with great social closures in social network, while men are more comfortable with competitive environments that come with many structural holes in social networks, and (2) the institutional reason that women collectively encounter gender discrimination, the so-called glass ceiling denotes the barriers and roadblocks to women's career trajectories that prevent them from reaching certain levels in corporate hierarchy. So, to break the glass ceiling, women need to bundle together to generate a strong and collective force.

5.2.2 *Online Discussion*

The literature on social network homophily provides a strong foundation for understanding which factors play a role in the formation of one's network ties but as advancements in technology provide us with more access to social media, networks are shifting online. The development of social networking sites allows for individuals, who may not be close in proximity or similar in demographics, to form social ties with each other. These sites can also facilitate network homophily due to the accessibility of information on one's status and values.

Regarding discussion of controversial social issues, one study reported that students felt more comfortable discussing controversial topics face-to-face and that younger people were more concerned with hurting feelings [22]. Additionally, semantic organization in online interfaces can lead to selective exposure to information on controversial topics [9]. These high-choice media environments allow people to familiarize themselves with like-minded messages, which shapes attitudes and perceived prevalence of opinions [27]. While some research has focused on the structure and features of negative interactions online (i.e. [28]), more studies are needed to examine whether homophily principle still underlies individuals' chatroom networking on controversial topics.

5.3 Research Questions

Synthesizing previous literatures, we identify two areas that are understudied: the distinction between positive and negative social networking and the invisible force of social networking. Homophily in social networking refers to conscientious selections of similar others in pursuing of common activities. A crucial presumption in social networking literature is that the network is formed for positive (or at least neutral) endeavors, such as a common interest, hobby, clubs, or fraternity initiation. During the process, social actors frequently use gender and racial biomarkers as the proxy for common interests and similar backgrounds, hence smooth communications, and less social anxieties. The underlying logic of this assumption is that the same gender and racial groups denote commonality and shared interests, conversely, different gender and racial groups represent dissimilarity in social backgrounds.

What about the social networking of negative purpose? Existing literatures are largely silent on this, especially in the literatures of social homophily. Does the same logic still operate? Do the same race and same gender still bound together even if it means negative connotations, such as gossiping networks, backstabbing among coworkers, or in our study, disputes, disagreements, and dismantling of each other's viewpoints on guns on campus? The lack of study on the subject of negative social networking is one of the main motivations for our research. Here, we can see clear demarcation between positive and negative chatroom networking on the issue of guns on campus. Irrespective of participants' pro-gun or anti-gun stance, participants in the positive chatroom network are agreeing and supportive of each other's stance, whereas the negative chatroom network are filled with disagreements and disputes while participants try to argue for their beliefs over the others.

Another area of understudy is the issue of visibility of identity of social network actors. Most social network research, especially in the literature of social homophily are based on data that network actors are fully aware of the demographics of their social contacts. However, what happens to homophily when gender and racial identity is not visible to network actors? Do the selections of similar others based on gender and racial similarity still take place? Addressing this issue is important because it reveals to what extent some underlying social forces play a role in social homophily above and beyond network actors' own cognitions. This is an area that is not well understood and is one of the focuses on our research. Using an innovation discourse platform allows us to garner participant's social demographic features, while hiding those features from the participants themselves, thus ensuring their chatroom networking is solely based on the chats.

5.4 Research Hypotheses

The extensive literature of homophily points us to the direction of same racial and gender selection in this chatroom networking on the controversial social issues of guns on campus. Filling in the gap of negative social networking and invisible selection of discussion partners in chatroom networking, our research should divulge to what extent the doctrines of homophily "birds with features fly together" works in this context of chatroom discussion among college students. In particular, our research sets up the stage to examine whether the homophily still operates when (1) the chatroom networking is negative, and (2) the selections of chatroom partners are made without the knowledge of the partners' racial and gender information. Here, in line with the general homophily literature, we hypothesize that,

H_1: Debates on the social issue of "guns on campus" are more likely to occur between the pairs of the same race than are pairs of different races across all chatroom networking (e.g., the overall chatroom, the positive chatroom, and the negative chatroom).

H_2: Debates on the social issue of "guns on campus" are more likely to occur between the pairs of the same gender than pairs of different genders across all chatroom networking (e.g., the overall chatroom, the positive chatroom, and the negative chatroom).

5.5 Data and Variables

We collect data from students enrolled in an "Introduction to Sociology" course during the spring of 2020, at a land-grant research university of a South East Conference (SEC) school. In particular, we used an online platform, named the Intelligent Cyber Argumentation System (ICAS), to facilitate online discussion among the students and collect their opinions posted on specific social issues. The ICAS platform is an experimental online deliberation platform that is designed to facilitate large-scale online discussion and debate and provide integrated analytical models to summarize and analyses the discussions automatically [18, 26]. The ICAS platform facilitates online discussions that are anonymous (users are assigned a randomly generated username at registration time), asynchronous (posts are persistent and visible to users at all times), and un-moderated (all posts are immediately visible to others after posting). ICAS is an issue-centric discussion platform, meaning that all of the discussions center around a topic issue.

This study overall focuses on the student's participation in online discussions using the ICAS platform. In the study, we asked the students to discuss and debate one particular controversial issue that was related to their campus life—guns on campus: should students with a concealed carry permit be allowed to carry guns on campus? The original raw data consists of user demographic information, which they enter at registration time, and online discussion threads, which users post during the discussions, contain thousands of online posts. We codify those threads with parsing languages, creating a dataset with key variables (gender, race, majors, family social economic status) and the whole discussion (all-edge) network depicting discussion interactions (reply-to posts) between dyads of participants. Out of the 125 total students who participated in the study, 75 participated in the ICAS online debate, producing 196 edges, or 2.6 edges per participant. In ICAS, when a user makes a post, they must explicitly state if they are agreeing or disagreeing with the post to which they are replying. Using this agree/disagree data, we further split the

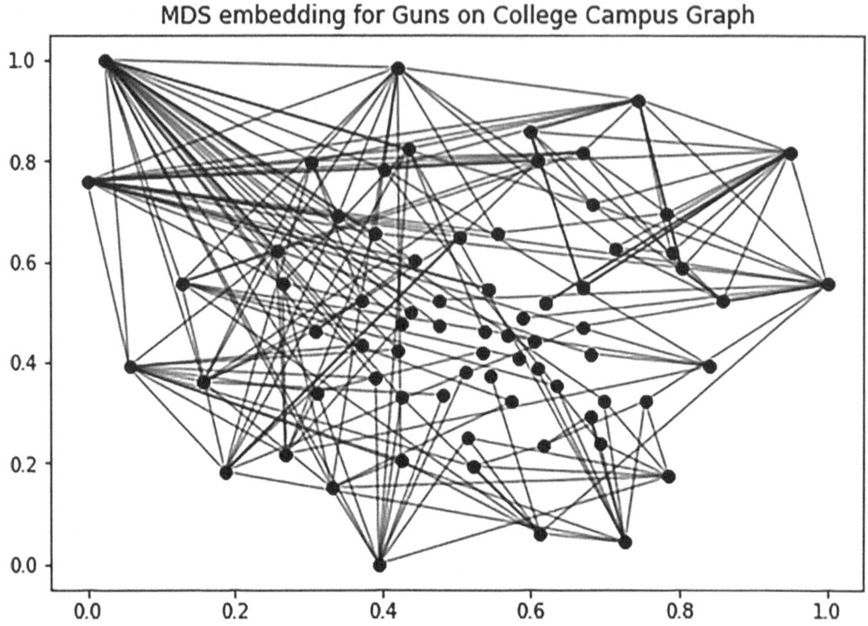

Fig. 5.2 MDS of the Chatroom Discussion Network (all edges)

network into two types: positive edge network, in which participants agreed with each other and reinforcing each other's viewpoints, and negative edge network, in which participants disagree, dispute, or try to dismantle each other's viewpoints on the issue. In total, 60 participants engaged in positive chatroom, producing 80 edges, or 1.3 edge per participant. For the negative chatroom discussions, 60 participants engaged in the debates, producing totally 95 edges, or 1.6 edges per participant.

Figures 5.2, 5.3, and 5.4 shows a visualization of all edge, positive edge, and negative edge networks using Multidimensional Scaling (MDS) embedding [14, 32]. It is immediately obvious from the visualization that one node (participant) has a disproportionately large number of edges in all edge and negative edge networks. We noticed some variations in the degree distribution for each of the networks. For all edge network, we see that most users have a degree of less than 10, with a few users having degree 10 or greater. For the positive edge network, we see that most of the users have a degree of less than 10, and only two users have a degree of 10 exactly. For the negative edge network, we see that most users have a degree of between one and five, with few users with degree six or more. One user in particular has a degree of 25, which is 26% of the total number of edges. Thus, in terms of degree, the negative network is more heavily skewed toward a small set of very active users, while the positive edge network has its edges more evenly distributed.

Table 5.1 shows the density, centralization, and transitivity for each of the discussion networks in the Guns on Campus issue. Overall, each of these graphs

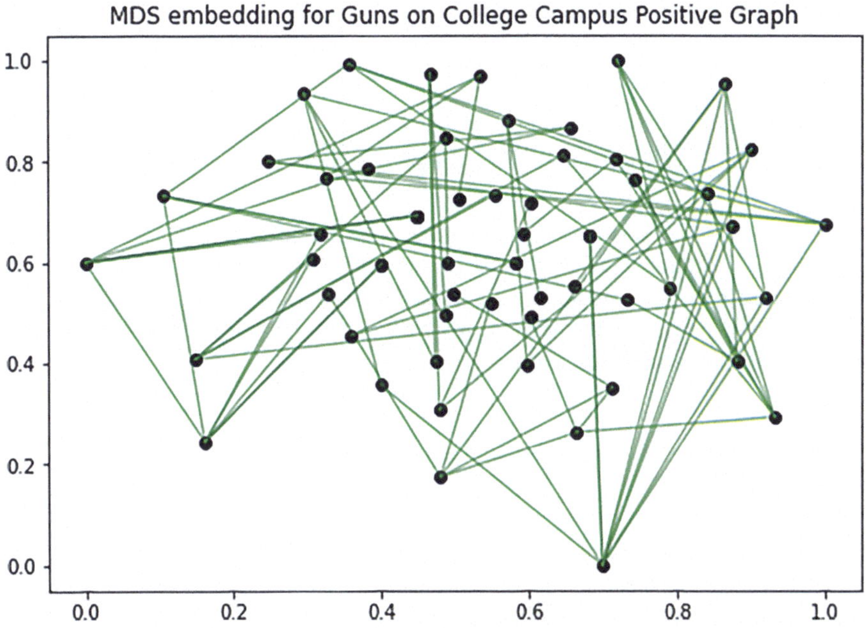

Fig. 5.3 MDS of the Chatroom Discussion Network (positive edges)

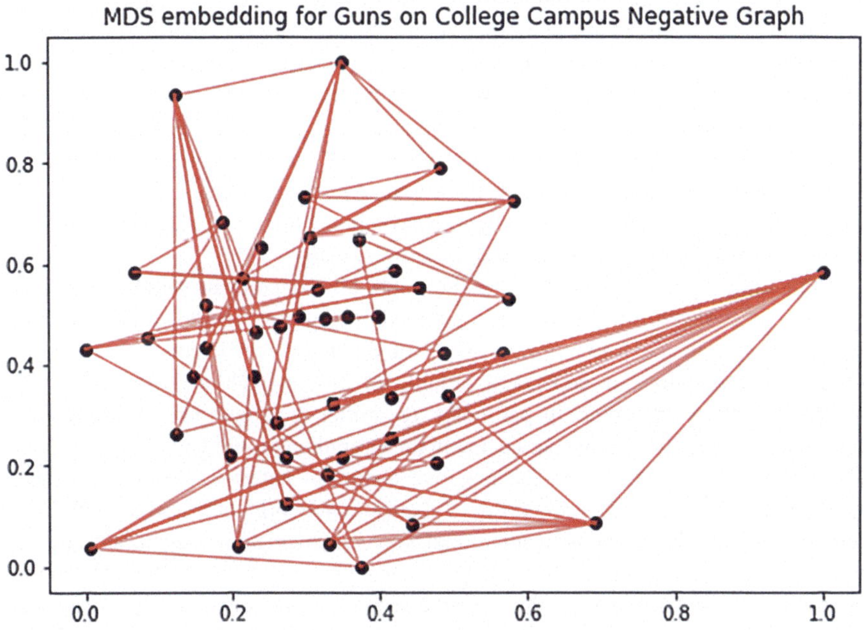

Fig. 5.4 MDS of the Chatroom Discussion Network (negative edges)

Table 5.1 The network measures of the three networks

	Density	Centralization	Transitivity
All edges network	0.0706	0.0039	0.1769
Positive edges network	0.0452	0.0022	0.0647
Negative edges network	0.0537	0.0065	0.0879

are very sparse with very low density and centralization scores. The negative edge network has a centralization value that is noticeably higher than all and the positive edge networks, which corresponds with our previous examination of degree distribution. The transitivity of the negative networks is also fairly low, indicating that the networks are sparse, and few cliques are forming.

5.6 Findings

We run separate regression models on the participation of chatroom of guns on campus. Table 5.2 shows the unstandardized coefficients of Exponential Random Graph Modeling (ERGM) of individual participation in the chatroom, which we ran using R's Statnet package. On this controversial issue of guns on campus, chatroom discussion network of all edges (all chatroom network) denotes the overall participation in the discussion. Chatroom discussion network of negative edges (negative chatroom network) indicates participation in chatroom disputes and disagreements. Chatroom discussion network of positive edges (positive chatroom network) suggests participation in chatroom agreements and mutual supports of viewpoints. In all models we report McFadden pseudo-R square, which uses the ratio of the fitted model's residuals to the null (intercept) model. The values reported show that the fitted models explain more than half the variance in the networks (value >0.5) and we can very strongly reject the null hypothesis that the fitted model does not explain any of the network ($p < 0.001$) based on our residuals.

Depending on the dependent variable graph, all independent variables, such as gender, race, college, and parental occupational prestige are significant factors of the participation in the chatroom discussions. Noted that the coefficients in ERGM of a binary undirected graph as dependent variable refer to the likelihood of presence of a network tie between a pair. In particular, regarding race, compared with Whites, Hispanics are less likely to be connected in the overall chatroom network, Asians are less likely to be connected in the negative chatroom network. Women are more likely to have connections in the positive chatroom network compared to men. Each unit of increase in parental occupational score also increases the likelihood of participating in the overall and negative chatroom network. Engineering students are more likely to participate in the overall chatroom network and negative chatroom network, compared with other majors such as architecture, arts, sciences, business, education, and health. In the positive chatroom network, significant difference emerged that agriculture majors are more likely to participate than are engineer

Table 5.2 Unstandardized coefficients of ERGM of Individual Participation in Chatroom Network with R Statnet

Single attributes	All edge network	Negative edge network	Positive edge network
Edges	−2.160*** (0.639)	−3.030** (1.113)	−1.806 (0.924)
Race: White	REF	REF	REF
Race: Black	−0.305 (0.406)	−0.683 (0.618)	−0.275 (0.541)
Race: Asian	−0.134 (0.196)	−0.708 * (0.322)	−0.027 (0.295)
Race: Hispanic	−0.558* (0.239)	−0.705 (0.497)	0.327 (0.327)
Gender: Male	REF	REF	REF
Gender: Female	0.146 (0.135)	−0.280 (0.199)	0.524** (0.201)
College: Agriculture	−0.437 (0.249)	−0.955** (0.337)	1.077* (0.466)
College: Architecture	−1.065** (0.329)	−2.439*** (0.553)	0.019 (0.599)
College: Arts and Sciences	−1.018*** (0.206)	−1.731*** (0.296)	−0.174 (0.367)
College: Business	−1.040*** (0.203)	−1.181*** (0.306)	−0.381 (0.348)
College: Education and Health	−1.228*** (0.224)	−1.544*** (0.316)	−0.493 (0.374)
College: Engineering	REF	REF	REF
College: Undeclared	−1.109* (0.433)	−1.222 (0.684)	−0.645 (0.599)
Average Parent's Prestige Scores	0.013* (0.006)	0.0316*** (0.009)	−0.015 (0.009)
Model R Square[a]	0.635***	0.730***	0.739***

* P < .05; ** P < .01; *** P < .001

[a]Reported McFadden Pseudo R Square for model fit

majors. The network endogenous variable under control is the number of edges in the network, which appear to be significant in overall and negative chatroom networks.

Perhaps the more revealing information is how the dyadic pair to pair matching affect the participation in the chatroom network of guns on campus. Table 5.3 shows the unstandardized coefficients of ERGM of chatroom participation with dyadic matching as independent variables. Racial disparity emerged in the participation of overall edge network and the positive edge network. In overall edge participation, interactions between white and Hispanics are significantly less likely to occur compared with interactions between two white students. In positive edge network, interactions between two Hispanic students are significantly more likely to occur compared with interactions between two white students. Such results partially support our first research hypothesis, with shades of gray. Racial homophily appears in pairs of white students in overall chat network, compared with interracial pairs of white and Hispanic students; it also appears in pairs of Hispanic students in positive edge network, compared with pairs of white students.

Table 5.3 Unstandardized coefficients of ERGM of Dyadic Participation in Chatroom Network with R Statnet

Dyad attributes	All edge network	Negative edge network	Positive edge network
Edges	−2.418*** (0.230)	−1.949*** (0.285)	−3.483*** (0.421)
Race: White—White	REF	REF	REF
Race: White—Black	−0.566 (0.470)	−0.372 (0.609)	−0.882 (0.736)
Race: White—Asian	−0.001 (0.203)	−0.270 (0.305)	−0.170 (0.316)
Race: White—Hispanic	−0.857** (0.280)	−0.909 (0.483)	−0.312 (0.394)
Race: Black—Black	NA	NA	NA
Race: Black—Asian	−0.091 (1.043)	NA	0.339 (1.054)
Race: Black—Hispanic	−0.069 (1.044)	NA	0.822 (1.071)
Race: Asian—Asian	−0.713 (1.031)	NA	−0.187 (1.045)
Race: Asian—Hispanic	−0.807 (0.729)	−0.381 (1.046)	−0.546 (1.030)
Race: Hispanic—Hispanic	0.079 (0.751)	NA	1.613* (0.816)
Gender: Male—Male	REF	REF	REF
Gender: Male—Female	−0.079 (0.234)	−0.607* (0.290)	0.252 (0.432)
Gender: Female—Female	−0.260 (0.246)	−1.510*** (0.338)	0.607 (0.436)
College—Matching	−0.217 (0.195)	−0.421 (0.292)	−0.582 (0.334)
Absolute Difference in Average Parent's Prestige Score	0.016 (0.008)	0.012 (0.013)	0.021 (0.012)
Model R Square[a]	0.625***	0.705***	0.735***

* $P < .05$; ** $P < .01$; *** $P < .001$;
[a]Reported McFadden pseudo–R Square for model fit

The most intriguing result appears in the gender matching in chatroom networking. First of all, gender matching does not exert any significant result for overall network or the positive edge network. Discussions pairs involving pairs of male students, female students, and male-female students do not affect the likelihood to participate in overall or positive edge network.

Gender matching appears to be highly significant in affecting the likelihood to participate in the negative edge network. Compared with the interactions between two male students, interactions between a male and a female student and between two female students are significantly less likely to occur in the negative edge network. Moreover, the likelihood that two women involved in disputes on the issue of gun on campus is the lowest among the three gender match configurations.

On the other side, arguments, disputes, and disagreements on the issue of guns on campus are the most likely to occur between two male students.

Such results are mixed in terms of supporting our second hypothesis. Gender homophily appears in pairs of two male students in negative edge network. When it comes to disputing and disagreeing with each other's viewpoints on guns on campus, such disputes are most likely to occur between two male students. However, contrary to the homophily literature, pairs of female-female students are the least likely to engage in those disputes, even compared with cross-gender pairs between a male and a female student. One of the obvious reasons could be that women are more ideologically aligned on this issue, so it would make sense that they would have fewer disputes. However, examining histogram data of the Pre-survey attitudes of the women and men, we found that while women have less ideological variance than men, they still have a significant amount of variance on this issue. So that explanation would likely not account for such a drastic difference between gender matching dyads.

Here, it is important to connect with the previous finding that at the individual level (see Table 5.2), women are more likely to participate in the positive edge network than are men, although the dyadic gender pairs do not show significant differences in the positive edge network (Table 5.3). We believe a much plausible explanation is that women simply shun away from negative comments, disputes, and disagreements over controversial issues.

5.7 Conclusions

While most literatures on social homophily focuses on interactions among similar individuals, this research leverages an innovative computer discussion platform to examine those hidden and invisible social forces that bound people together. The platform is so designed that students logging into the system all have randomly assigned screen names that they do not know each other's identity. Meanwhile, we collect the basic demographic information, as well as the digital traces of their chatroom behaviors (with whom they discuss, and the contents of their discussions). We process and encode the user demography data and discussion interactions to produce discussion matrices for our network analyses.

So, the chatroom networking activities captured in our analyses are truly the candid shots of the participants who are unaware of the social identities of their conversation partners. In addition, we also contextualize our chatroom networks by separating between positive edge network and negative edge network. Although network literature overwhelming focuses or presumes that network inactions are positive, many network interactions are negative. In our particular research, gun on campus is a very controversial issue, that draws conflicting views among students. In our positive chatroom network, students are mutually supportive of each other's views, which reinforces the beliefs from network participants (pro or con of guns on campus) on the issue. In contrast, the negative chatroom network is filled with conflicting viewpoints, resulting in eruptions of disputes and disagreements among participants. Structurally, the negative chatroom network also has the highest centralization, which indicates that the extent to which the differences between network actor in their centrality are high: a few opinionated individuals dominate the room by posting many times, while most others are less active with fewer discussion posts.

Our central contributions to the literature of homophily rests in our contextualized analyses of online chatroom networking behaviors in an identity blind environment and across different chatroom tunes (the positive edge and negative edge network). The key research question we ask here is "do the racial and gender homophily still operate in negative networking and in a network whereby participants cannot see each other's racial and gender identities." Our findings provide mixed results in lending their support to the racial and gender homophily. In the overall network (both positive and negative chatroom together), pairs of white students are most likely to form, and are significantly more likely to materialize than cross racial pairs between a white and a Hispanic student. Racial homophily does not appear in other chatrooms.

Gender homophily only appears in pairs of two male students in the negative edge network. Pairs of two female students are the least likely to participate in the negative edge network, contradictory to the gender homophily literature. This, along with the individual level analyses that female students are more likely to engage in positive edge networking than are male students, aligns with Burt's [5] discovery of gender disparity in capitalizing structural hole benefits. Unlike men, women in general disengage from the highly competitive environments. Instead, women tend to thrive in a nurturing, and mutually supportive context. Burt offered two explanations: the cultural explanation that women are raised in a more nurturing and supportive environment, whereas men are raised in more combative setting. The institutional explanation that women need to bundle together to break the glass ceiling that impedes their career development in professional workplaces. Because our research took place in University classrooms, our finding should allude to the cultural explanation more than to the institutional explanation.

Our research aims at the online virtual engagement of a controversial social issue of guns on campus. In such context, we investigate the issue of homophily along gender and racial dimensions. Essentially, this is an interest group homophily analysis, examining the roles of gender and race in attracting students to the online discourse, either agreeing or disagreeing with each other on their respective pro or anti-gun stances. One line of future research can examine the in person real-life interactions of some interest groups or debate clubs. Significant differences may exist between virtual debate and real-life in person debate: the virtual debate is driven by its laser-focus on the threads posted, drawing a crowd of like-minded participants. In contrast, in-person debate may have multiple focuses at a time, drawing participants into different directions. A study of in-person debate can provide a comparison, thus contextualizing our study in the literature of homophily.

Our data are based on a class of students, who are age and educational homophilous. In reality, age and education can be important forces in the literature of social homophily. For example, age homophily is rather low in American's core discussion group, which remains almost unchanged from 1985 to 2004 [21]. In contrast, educational homophily is fairly high in American's core discussion group, which has been increasingly from 1985 to 2004. In other words, while American's core discussion groups involve persons across diverse age groups, those groups largely contain persons with similar educational backgrounds. Such features do not change much from 1985 to 2004. The key message from this line of research is that we need to account for the age and education in studying of homophily, which is not the case in our research, but an important task for future studies.

References

1. Aristotle (1934) Rhetoric. Nichomachean ethics. In: Aristotle in 23 Volumns Rackman transl. Harvard University Press, Cambridge
2. Bott H (1928) Observation of play activities in a nursery school. Genet Psychol Monogr 4: 44–88
3. Burt R (1992) Structural holes. Harvard University Press, Cambridge, MA
4. Burt R (1997) The contingent value of social capital. Adm Sci Q 42(2):339–365. https://doi.org/10.2307/2393923
5. Burt RS (1998) The gender of social capital. Ration Soc 10(1):5–46. https://doi.org/10.1177/104346398010001001
6. Duncan OD, Featherman DL, Duncan B (1972) Sociometric background and achievement. Seminar, New York
7. Eder D, Hallinan MT (1978) Sex differences in children's friendships. Am Sociol Rev 43: 237–250
8. Fischer CS (1982) To dwell among friends. University Chicago Press, Chicago
9. Gao M, Do HJ, Fu W-T (2017) An intelligent interface for organizing online opinions on controversial topics. In ACM 22nd annual meeting of the intelligent user interface, March 13–16
10. Hallinan MT, Smith SS (1985) The effects of classroom racial composition on students' interracial friendliness. Soc Psychol Q 48:3–16
11. Hubbard RM (1929) A method of studying spontaneous group formation. In: Thomas DS (ed) Some new techniques for studying social behavior. Child Development Monographs, pp 76–85
12. Ibarra H (1995) Race, opportunity and diversity of social circles in managerial networks. Acad Manag Rev 38:673–703
13. Kalmijn M (1998) Intermarriage and homogamy: causes, patterns, and trends. Annu Rev Sociol 24:395–421
14. Knoke D, Yang S (2020) Social network analysis. SAGE Publications, Inc., Thousand Oaks
15. Laumann EO (1973) Bonds of pluralism: the form and substance of urban social networks. Wiley, New York
16. Lazarsfeld PF, Merton RK (1954) Friendship as a social process: a substantive and methodological analysis. In: Berger M (ed) Freedom and control in modern society. Van Nostrand, New York, pp 18–66
17. Lincoln JR, Miller J (1979) Work and friendship ties in organizations: a comparative analysis of relational networks. Adm Sci Q 24:181–199
18. Liu XF, Khudkhudia E, Wen L, Sajja V, Leu MC (2010) An intelligent computational argumentation system for supporting collaborative software development decision making. In: Meziane F, Vadera S (eds) Artificial intelligence applications for improved software engineering development: new prospects. IGI Global, pp 167–180. https://doi.org/10.4018/978-1-60566-758-4

19. Loomis CP (1946) Political and occupational cleavages in a Hanoverian Village. Sociometry 9: 316–333
20. McPherson M, Smith-Lovin L, Cook JM (2001) Birds of a feather: Homophily in social networks. Annu Rev Sociol 27(1):415–444
21. McPherson M, Smith-Lovin L, Brashears ME (2006) Social isolation in America: changes in Core discussion networks over two decades. Am Sociol Rev 71(3):353–375
22. Meyer KA (2006) When topics are controversial: is it better to discuss them face-to-face or online? Innov High Educ 31:175–186
23. Nooraie RY, Wayman E, Dozier A (2020) Gender Homophily in translational collaborations; a network analysis study of investigators at one Academic Medical Center. J Clin Transl Res 4(1):118
24. Shrum W, Cheek NH, Hunter SM (1988) Friendship in school: gender and racial Homophily. Sociol Educ 61:227–239
25. Simmel G, Levine DN (1971) On Individuality and social forms: selected writings. University of Chicago Press, Chicago
26. Sirrianni J, Liu X, Adams D (2020) Agreement prediction of arguments in cyber argumentation for detecting stance polarity and intensity. In: Proceedings of the 58th annual meeting of the Association for Computational Linguistics, pp 5746–5758. https://doi.org/10.18653/v1/2020.acl-main.509
27. Sude DJ, Knoblock-Westerwick S, Robinson MJ, Westerwick A (2019) "Pick and choose" opion climate: how browsing of political messages shapes public opinion perceptions and attitudes. Commun Monogr 86(4):457–478
28. Tsvetkova M, Garcia-Gavilanes R, Yasseri T (2016) Dynamics of disagreement: large-scale temporal network analysis reveals negatvie interactions in online collaboration. Nat Sci Rep 6:36333. https://doi.org/10.1038/srep36333
29. Verbrugge LM (1977) The structure of adult friendship choices. Soc Forces 56:576–597
30. Wellman B (1929) The school child's choice of companions. J Educ Res 14:126–132
31. Yang S, Jackson BA, Zajicek A (2021) A changing landscape? An intersectional analysis of race and gender disparity in access to social capital. Sociol Spectr 41(1):80–95
32. Yang S, Keller FB, Zheng L (2016) Social network analysis: methods and examples. Sage, Thousand Oaks
33. Zelter D (2020) Gender homophily in referral networks: consequences for the medicare physician earnings gap. Am Econ J Appl Econ 12(2):169–197

Chapter 6
An Ego-Centric Network Study of Core Discussion Network During the Covid-19 Pandemic

Song Yang, Douglas Adams, and Regan Hodges

Abstract This chapter uses data mined from ICAS to conduct ego-centric network analysis of students' core discussion network. In particular, the ICAS recorded online discourses on five social issues such as school closing, facemasks, collegiate sports, race and policing, and social inequality from students taking a General Sociology class during the Fall 2020, when the class was held online. For the total 191 students who participated in the discussions, we elicit 847 conversation partners (alters) with our five social issue generators, producing an average of 4.43 alters per respondent. The core discussion groups of our sample are very gender balanced, with female alters slightly outnumbering male alters. However, the core discussion groups are racially homogenous, consisting of mostly white students (around or above 80 percent). Explanatory analyses reveal that gender and race of respondents significantly impact the size, gender composition, and racial composition of their core discussion networks. We discuss those major findings and implications of future studies in our conclusion section.

Keywords Ego centric network analysis · Core discussion groups · Social issue generators · Covid-19 discussion networks

This chapter uses data mined from ICAS to conduct ego-centric network analysis of students' core discussion network. In particular, the ICAS recorded online discourses on five social issues such as school closing, facemasks, collegiate sports, race and policing, and social inequality from students taking a General Sociology class during the Fall 2020, when the class was held online. For the total 191 students who participated in the discussions, we elicit 847 conversation partners (alters) with our five social issue generators, producing an average of 4.43 alters per respondent.

S. Yang · D. Adams
Department of Sociology and Criminology, University of Arkansas, Fayetteville, AR, USA
e-mail: yangwang@uark.edu; djadams@uark.edu

R. Hodges (✉)
Simmons Foods, Rogers, AR, USA
e-mail: reganharper98@gmail.com

S. Yang (ed.), *Social Network Analysis in Action*, Lecture Notes in Social Networks,
https://doi.org/10.1007/978-3-031-66661-2_6

The core discussion groups of our sample are very gender balanced, with female alters slightly outnumbering male alters. However, the core discussion groups are racially homogenous, consisting of mostly white students (around or above 80 percent). Explanatory analyses reveal that gender and race of respondents significantly impact the size, gender composition, and racial composition of their core discussion networks. We discuss those major findings and implications of future studies in our conclusion section.

6.1 Introduction

The historically unprecedented Covid-19 has shutdown campus, closing offices, gyms, shopping centers, overwhelming clinics and hospitals. Draconian measures such as shelter in place has locked people at home, isolated and disconnected from each other. Social distancing become the word of the year or even of the century as one of the most important social mechanisms to fight this highly transmissible disease. However, during this challenging time, social networking and connectivity between each other, which is the polar opposite of the social distancing has never been such critically important until now. We as individuals are social beings craving for connections with others. In the past we rely on many social settings, such as schools, workplaces, gyms, coffee stores, shopping center to construct and jump start our social networking and social connectivity. As the pandemic shut those down those venues one after another, how to maintain social connectivity and avoid social isolation? Do people still engage in social networking and establish connections with each other? To those questions, this research seeks answers by surveying college students enrolled in an Introduction to Sociology with our innovative social issue generators.

Our research situates closely with literatures on ego-centric network studies, mostly represented by the two rounds of General Social Surveys (GSS) of "core discussion group" in 1985 [31] and 2004 [33]. The study of the core discussion group in 1980s [31] revealed that American core discussion groups were kin-centered, racially homogenous and gender balanced. Fast forwarding 19 years later in 2004, those groups become more kin-centered, racially homogenous and gender balanced than they were in the 1980s [33]. Consequentially, Americans' core discussion groups shrink, with many individuals suffer from total isolation, having no one to discuss with on important matters.

Understandable, such finding ignites many debates among scholars as well as popular press. Some confirm the finding of social isolation and called for greater effort to mobilize civic and community engagements. Others questioned the validity of the finding, suggesting possible sources of bias such as wording of the questionnaire items, sequencing of the items in the entire interview process, interviewer training, and interviewee fatigues. As the debates of social isolation seems to calm down in recent years, the arrival of Covid-19 pandemic reignites the literature of isolation and loneliness, with great level of attention to the comparison of online versus offline and in person social networking and connectivity.

Under such context, we re-examine the issue of social networking and interpersonal connectivity using an alternative questionnaire item prob: social issue generators. We surveyed college students with this new prob. during the Fall of 2020, right in the middle of the raging pandemic. Sections below discuss the literatures on important matter core discussion network, our rationales for developing an alternative prob. for surveying individuals' social networks, our research questions and hypotheses, data and variable constructions, findings, and conclusion.

6.2 Literature on Core Discussion Group

When discussing important topics, such as the Covid-19 pandemic or race relations between the community and police, it is typically with individuals who are considered close. These individuals, or 'discussion partners,' can consist of intimate partners, friends, neighbors, co-workers, family, and more. The Covid-19 pandemic of 2020 resulted in a worldwide shutdown of business and institutions and people were asked to stay home. The outcomes of this event remain to be evaluated and the current study aims to understand how social networks are formed in relation to discussing matters like Covid-19 and the racial relations that swept the nation. In the following sections, we will provide a brief history of the social network name generator and follow up with some differences observed in social networks based on demographic characteristics, like race and age. We then present our research using social issue generators to gauge individuals' social networking activities during this unparallel time of pandemic that wrack havoc social connectivity.

6.2.1 Social Network Studies with the Name Generator

For decades researchers have been interested in how individuals form their social networks [25, 53]. Researchers developed many instruments used to obtain survey measures of networks (e.g., [11, 16, 29, 49]), while Burt's [9] name generator is one of the early tools for eliciting network data for ego-centric network. In particular, the 1985 General Social Survey (GSS) used the name generator to ask respondents, "from time to time, most people discuss important matters with other people. Looking back over the last 6 months – who are the people with whom you discussed matters important to you?" The survey also asked respondents to provide information on their discussion partners' demographic characteristics such as sex, race, education, age, and the type of relationship they have. Later research shows that when asked this question, respondents typically list people with whom their contact is most frequent and people who are seen as close ties tend to be listed first [8, 50]. Having close ties to someone reflects the amount of support they are willing to provide [51] and increases the likelihood that they would provide help during a crisis [23].

The ability to elicit names using a name generator provides researchers with more detailed information on the respondent's, or *ego*'s, close ties. Such ego-centric network study reveals the relationships from the point of view of a single individual. In addition, GSS is a national representative sample, and findings from the ego-centric study is readily applicable to the US population. The studies using the 1985 GSS social network data focuses on a variety of topics, such as kinship [35], network range and its association to socioeconomic status [20], structural features relating to political behaviors [24, 46], co-worker networks [21, 46], differences based on marital status [1, 22, 28], subcultural differences [5, 13]; and how heterogeneity reinforces attitudes [6]. The name generator succeeds at eliciting an individual's 'core' discussion network [3].

Almost 20 years later, McPherson et al. [33] analyzed the ego-centric network data in the 2004 General Social Survey (GSS), which used the same name generator as the 1985 GSS. They found that the average size of Americans' core discussion networks had shrunk by about a third from 1985 to 2004, and the number of respondents who did not report any friends had increased almost twice [31, 33, 34].

Their analyses showed that the percentage of respondents who reported nobody to discuss important matters with increased from 10.0 percent in 1985 to 24.6 percent in 2004. In addition, McPherson et al. [33] also reported that both the number of kin confidants and the number of non-kin confidants decreased during the years from 1985 to 2004 [33, 34]. The percentage of respondents who reported at least one kin confidant decreased from about 70 percent in 1985 to about percent in 2004, and the percentage of respondents who reported at least one non-kin confidant decreased from about 64 percent to about 54 percent. Based on these findings, they concluded that Americans are much more isolated than they were in 1985.

McPherson [33] et al.'s findings attracted tremendous public attention, as well as a great deal of scholarly debate [14, 15, 17, 33, 34, 37, 38]. The conclusion of decreasing social connection or increasing social isolation drew widespread coverage in public media, and people believed that Americans were more isolated and became more anxious about the lonely Americans [17]. Their findings correlate with Putnam's [40] assertion that Americans experienced the declining civic engagement and neighborhood involvement in the last 50 years. The less participation undermines individual's connections within communities and makes people become more isolated [40]. The findings also correspondent with the observation of *The Lonely America* [36] that online activities reduced physical relationships, decreasing individual's social capital and increasing social isolation. Other researchers similarly stated that the new social networking site tend to contribute to the decline of social interaction, increasing social isolation [7, 43]. Indeed, the biggest irony in the social networking world of the twenty-first century is that those social networking platforms that strive to keep us connected end up setting us apart.

Nevertheless, other scholars delve deep into the data and other evidence are not quickly convinced of the social disconnection and increasing isolation of Americans. Researchers call McPherson et al. [33] findings an artifact, resulting from a number of issues from study design, to interview effect, fatigue, and a random technical error of the GSS [14, 37]. For example, the section of network questions in the 2004 GSS came late in the whole interview, after the complicated questions about organizational membership. This design made respondents feel tired and reluctant to report more confidants. As a result, the reported number of confidants could be lower than the real number. In addition, interviewers used different actions, probes, and interactions during an interview. These variations during an interview could also result in some errors. The training and interview effects are significant contributing factors to increased errors in the 2004 GSS. Researchers also noted some technical errors or computer errors during the interview and coding process [37]. In total, these random errors resulted in at least 15 percent to 20 percent of respondents who are member of subgroups to be coded as isolated [14]. All those findings cast ambiguities to the ego-centric network data of the 2004 GSS, producing lots of doubts on the findings of American's increasing isolation in the 2004 GSS.

6.2.2 The Meanings of Important Matters

While the name generator is critical in generating respondents' ego-centric network as it was applied to two rounds of national representative surveys. The design of the questionnaire item leaves it open for respondents to interpret what those important matters mean to them. So, what are those important matters and what affect respondents' conceptualization of those important matters? It turns out that a significant variation exists in the interpretation of the important matters, while many respondents considered "important matters" as personal or intimate relationships or personal life, and others think those more align with work and political discussions [3].

Researchers also discovered that the interpretation of important matter is vulnerable to the probe, a set of questions imposed right before the important matter questions [3]. Sometimes, interpretations of important matters correlate with the major societal affairs. For example, when respondents are questioned during a season of heated discussion about war and presidential primaries, the "important matters" tend to be interpreted as political matters, contributing to the increase of network size [14]. Such contextual effects also exhibit across different cultural and institutional environments. In 1980s, Chinese tend to discuss important matters, interpreted as personal issues with their coworkers. Such inclusion of coworkers as personal confidants to discuss important personal issues forms stark contrast to the core discussion networks in the west. Some 20 years later, those important matters discussion partners change to personal friends and family members [41], which is consistent with findings of core discussion partners in the west. Ruan [41] observed such change reflects the transition from command economy in 1980s that centralized work units to market economy in late 1990s in China, which caused decline of workplaces in individuals' social life.

6.3 Social Issue Generators: An Alternative Approach

The name generator premises on the important matters to respondents, which are subject to diverse interpretations. As American society collectively experiences great diversity and more polarized views on social affairs, we state that a more straightforward prob. to examine Americans social connections is in order. This is especially the case in this unprecedent time of global Covid-19 pandemic that wreak havoc many existing social orders, setting apart different social groups with different social and political viewpoints. The important matter prob. needs an update, and we feel our "social issue generator" can be better equipped to address the challenges of examining personal networks and individuals' social connections in current turbulent times.

The most important change between our social issue generators and the important matter generator is that the social issue generators do not leave much room for respondent interpretation. Rather than asking respondents "with whom they discuss important matters," we asked respondents "with whom they discuss important social issues," such as Covid-19 and facemasks etc. Such design pinpoints the exact social issues that interests us, providing valuable data to examine individual social connections revolving around important social issues. Another benefit of using social issue generators is that researchers can change those social issues, depending on what those most important social issues are in current society. Such feature of changeability is highly desirable given that the important social issues change much quicker during the unmatched time catalyzed by the covid-19 pandemic.

To reflect current social affairs for this research, we developed five social issue generators, such as covid-19 and school closings, covid-19 and facemasks, covid-19 and collegiate sports, race and economic inequality, and race and police. We surveyed college students with those social issue generators with an innovative computer platform that captures online discourse networks. In the following sections, we discuss our research questions, hypotheses, data sources, variables, findings, and draw some conclusions.

6.4 Research Question and Hypotheses

Our research using our five innovative social issue generators investigates two research questions: (1) the size of their social networks, and (2) the composition of their social network. Both research topics are critically important during the unique covid-19 pandemic. The first research question is related to the debate of social isolation over the last few decades. As previously noted, researchers [33] reported a steady decline of American's social networking over the 20 years from 1985 to 2004. Their findings inspired waves of academic debates as well as popular press coverage. Are Americans become more socially isolated and lonely? Or their findings are simply artifacts resulting from random issues such as "important matter prob" design issues, interviewer training, interview fatigue, or sequencing of the questionnaire items.

While the debates have no end in sight, concerns of social isolation and lonely abound. Researchers focus on the consequences of social isolation, documenting social capitals deficiencies as well as serious health consequences as results of loneliness. For example, social isolation leads to increased morbidity and mortality [10, 42, 48], infection [12, 39], depression [19], and cognitive decline [4, 52]. Scholars report those factors often go hand in hand, such as having small social networks, low participation in social activities, perceived lack of social support, and feelings of loneliness [18, 47]. The advent of and ongoing covid-19 pandemic pushes the issues of social isolation and loneliness to a new height, causing social scientists and public health officers to call for scholarly investigations of antecedences and consequences of social isolation [26]. Under such context, we examine the critical aspect of social networking: the network size with a survey of college students using our innovative social issue generators. Due to our social issue generator that is brand-new and the unparalleled time of covid-19 pandemic, we are not proposing any specific hypotheses here, but rather investigate this as a general research topic: does the size of social discussion network decline in the middle of the pandemic.

Our second research focus is the network composition of the social issue discussion networks. Here, sociology literatures accumulate large body of knowledge of social homophily in social networking, tracing back decades ago when Lazarsfeld and Merton [27] and Simmel's [44] used the term "homophily" and "birds of a feather flock together" to describe the tendency of being similar in individuals' social networks. The tenet of social homophily in social networking is that similar people bound together. While similarity breeds connections, what are those similarities? To this, McPherson, Smith-Lovin, and Cook [32] provided a thorough examination of the similarities and homophily of social networking. They documented extensive social homophily along multidimensional such as gender, race, age, education, social class, occupation, social attitudes, and behaviors. They also delved deep into the factors behind the force of social isomorphism, discussing the influences of occupations, families/kinship, work organizations, and geography.

Both rounds of GSS core discussion group reported well balanced gender composition, but very tilted racial composition in Americans' core discussion network. Scholars ([31, 33]) attribute such high gender heterogeneity and low racial heterogeneity to the importance of kinship in core discussion networks. In 1980s, Americans named many families and relatives in their core discussion networks. In 2004, they nominated even more families, especially core nuclear family members (spouses and parents) in their core discuss groups. Consequentially, the core discussion network encompasses a balanced number of men and women, but an uneven number of racial groups (majority of families belong to one race).

Such argument prompts us to further examine this complex issue. For example, if kinship is the main factor driving the high heterogeneity of gender and low heterogeneity of race, does kinship affect men and women differently in terms of gender homophily in social networking? Does kinship affect different racial groups differently in terms of racial homophily in social networking? Those issues beg serious investigations because although gender gap in network size (women have larger network than do men) remain unchanged between 1985 and 2004, their respectively network compositions change. American women have smaller kinship ties and large outside kinship contacts in 2004 than they were in 1985 [33]. However, very few study exists to examine the complex interactions between kinship and gender preference in naming similar others in their respective social network. For example, while gender preferences may present in that men name other men, and women name other women in their discussion networks, how such preference change (or unchange) with increases of kin ties in discussion networks?

On racial homophily, scholars discussed the differentials of racial homophily between whites and minorities—while all have tendency to include others in their networks who are from the same race group as the egos, such propinquity is stronger among whites than it is among minorities, causing racial gaps between whites and minorities in racial homophily of their networks. In addition, while researchers asserted that the main force of racial homophily in discussion network is the kinship ties—family ties are relations with other family members who are most likely to be the same race with the egos, very little is known about how kinship ties affect different racial groups differently its impact on racial homophily of their networks.

Along the gender homophily, we propose that

H_1: gender homophily appears in social issue discussion groups: compared with men, women's discussion networks have more women discussion partners.

While both men and women include significant number of kinships in their discussion networks, how the nomination of family members affect the gender homophily between men and women. Previous literatures are largely silent on this issue, but we believe that gender homophily goes beyond the family boundary. That is: women include other women in their discussion groups not through family ties, but through social setting beyond the realm of families.

H_2: A significant gender gap appears in how family ties affect the inclusion of women in social issue discussion networks. While family ties increase the inclusion of other women in men's social networks, family ties decrease the inclusion of other women in women's social networks.

Along the racial homophily, we propose that

H_3: Racial gaps appear in racial homophily, that whites have more racial homophily compared with other minorities in their social discussion networks.

In regarding the effect of kinship on racial homophily, researchers attribute the kinship as the main contributor to the racial homophily in core discussion networks [33]. However, scant research exists to investigate how such force of racial isomorphisms due to family ties in social networking can be different between different racial groups. This is perhaps due to a common sensical approach that if one includes others in the family in its social networks, the racial homophily would increases because the others are mostly likely to be the same race as the ego, regardless of which racial groups the egos belong. However, we think such simply assumption needs to be revisited, because American families maybe more diversified than it was. Here, we speculate that while family ties are racial homophilous, such trend is more obvious for white families than it is for minority families. Hence,

H_4: Racial gaps appear in how family/kinship ties affect racial homophily in social discussion networks, the effect of the inclusions of family ties on increase of racial homophily in one's social network is more pronounced for whites than it is for minorities.

6.5 Data and Variables

We collect data from students enrolled in an "Introduction to Sociology" course during the Fall of 2020, at a land-grant research university of a Southeast Conference (SEC) school. In particular, we used an online platform, named the Intelligent Cyber Argumentation System (ICAS), to facilitate online discussion among the students and collect their opinions posted on specific social issues. The ICAS platform is an experimental online deliberation platform that is designed to facilitate large-scale online discussion and debate and provide integrated analytical models to summarize and analyses the discussions automatically [30, 45]. The ICAS platform facilitates online discussions that are anonymous (users are assigned a randomly generated username at registration time), asynchronous (posts are persistent and visible to users at all times), and un-moderated (all posts are immediately visible to others after posting). ICAS is an issue-centric discussion platform, meaning that all of the discussions center around a topic issue.

The study design uses ego-centric network framework, which was used in two previous rounds of General Social Survey social network studies [31, 33]. The questionnaire instrument used to capture such ego-centric personal network is social issue generator, which reads "Occasionally, people discussion important issues with other people. Looking back over the last six months, who are the people with whom you discussed the following important topics: 1) Covid-19: Closing Schools, or not. 2) Covid-19: Wearing Masks, or not. 3) Covid-19: Getting Tested for Covid-19, or not. 4) Race: and Economic Inequality, and 5) Race: and the Police." Students can name up to five social contacts (alters) with whom they discuss one or multiple topics listed above. For each alter named by the students, we ask about its gender, race, and types of relation (kinship, spouse/partners, friends, and acquittances) with the students. We have totally 191 students completed the online survey, out of 472 the total number of students in the class, achieving a 40 percent response rate. Most of our respondents are young ($\overline{X} = 18.76$), white (80.6%), and female (71.2%). The majors of their studies are very diverse, across different colleges (see Appendix).

6.6 Findings

To address our first research question, are Americans become more isolated and lonelier (e.g., the size of social issue discussion network)? It does not appear to be the case, at least with our college student sample. Table 6.1 shows that, altogether, our 191 respondents nominated 847 alters with whom they discuss the five Covid related social issues, producing an average of 4.43 alters per respondents. The mode of the social issue discussion network size is 5, as most of the respondents (74.9% or 143) in our sample named all five alters with whom they discuss those issues. Those numbers are higher than those statistics in 1985 GSS ($\overline{X} = 2.94$; Mode = 3) and 2004 GSS ($\overline{X} = 2.08$; Mode = 0). One obvious reason is that we use five social issues to elicit alters, whereas GSS 1985 and 2004 module used one single name generator to elicit alters. Another reason is perhaps related to our sample being college students, who are immersed in the campus culture that encourages debates, discussions, and social networking. If it was not due to Covid-19 that resulted in campus shutdown during this survey, it would have been more responses to our social issue generator items. Simply put, when it comes to finding conversation partners to discuss current critical social issues, people have no problem finding those others.

Before we delve deep into the explanatory analyses of gender and racial heterogeneity, we would like to offer some preliminary analyses of the descriptive statistics of the ego-centric network of our social issue discussion networks. Table 6.2 shows the composition along three important dimensions: relational type, gender, and race. We separate those compositions by the size of the discussion networks. Looking at the type of relationship, for those whose networks have at least one partner, families, or relatives account for the majority of discussion group. However, the proportions of friends increase with the increase of network size. For those whose networks have at least two partners, friends account for 38 percent of the network members. That number keeps increasing with the increase of network size. Friends are the majority, account for 49 percent of network partners for those whose networks have five members. Such finding suggests that people first think of their family members when asked about with whom they discuss the current social issues. Such familial

Table 6.1 Frequency distribution of egos and alters nominated

Number of alters named	frequency	Cumulative frequency
5	143	143
4	14	157
3	16	173
2	8	181
1	9	190
0	1	191
N	191	–

All together, we have 847 names nominated by 191 respondents, producing an average of 4.43 nominations per respondents

Table 6.2 Descriptive statistics of the five alters

Variables	*Network* ≥ 1	*Network* ≥ 2	*Network* ≥ 3	*Network* ≥ 4	*Network* = 5
Relationship	1: 87 (45.8%)	1: 87 (48.1%)	1: 79 (45.7%)	1: 64 (40.8%)	1: 51 (35.7%)
1: Families/relatives	2: 40 (21.1%)	2: 6 (3.3%)	2: 6 (3.5%)	2: 6 (3.8%)	2: 6 (4.2%)
2: Spouses/partners	3: 52 (27.4%)	3: 69 (38.1%)	3: 72 (41.6%)	3: 77 (49.0%)	3: 70 (49.0%)
3: Friends	4: 11 (5.8%)	4: 19 (10.5%)	4: 16 (9.2%)	4: 10 (6.4%)	4: 16 (11.2%)
4: Acquittances	IQV = 88.97%	IQV = 81.52%	IQV = 81.12%	IQV = 78.39%	IQV = 82.42%
Gender	1: 79 (41.6%)	1: 70 (38.7%)	1: 67 (38.7%)	1: 64 (40.8%)	1: 56 (39.2%)
1: Men	2: 111 (58.4%)	2: 111 (61.3%)	2: 106 (61.3%)	2: 93 (59.2%)	2: 87 (60.8%)
2: Women	IQV = 97.18%	IQV = 94.89%	IQV = 94.89%	IQV = 96.61%	IQV = 95.33%
Race	1: 154 (81.1%)	1: 141 (77.9%)	1: 137 (79.2%)	1: 127 (80.9%)	1: 118 (82.5%)
1: White	2: 9 (4.7%)	2: 14 (7.7%)	2: 10 (5.8%)	2: 10 (6.4%)	2: 11 (7.7%)
2: Black	3: 14 (7.4%)	3: 12 (6.6%)	3: 13 (7.5%)	3: 11 (7.0%)	3: 10 (7.0%)
3: Hispanics	4: 13 (6.8%)	4: 14 (7.7%)	4: 13 (7.5%)	4: 9 (5.7%)	4: 4 (2.8%)
4: Others	IQV = 43.00%	IQV = 50.26%	IQV = 47.75%	IQV = 44.44%	IQV = 41.03%
N	190	181	173	157	143

boundary is pushed outward to include friends when they include more people in their discussion networks.

Gender composition is also very well balanced, with most conversation partners being female, corresponding to the most respondents being female. Comparatively, racial composition is much more tilted, concentrating on mostly white being included as conversation partners. In particular, around 80 percent of those in the network are white, which correspondence to our sample with mostly white students. Such results provide preliminary evidence that gender and racial homophily may play a role in forming those discussion networks.

To analyze the gender and racial composition of the students' social issue discussion groups, we regress female proportion and racial IQV (Index of Qualitative Variables) on students' age, majors, network sizes, kin proportion, gender, race, and the multiplicative terms between kin proportion and gender, and between kin proportion and race. The average female proportion in the social issue discussion network is 59.8 percent, with standard deviation of 25.3 percent. Those numbers for racial IQV are 24.5 percent and 29.3 percent respectively. One methodology note worth mentioning is that to avoid multicollinearity between the main terms and their product terms, we centered the kin proportion (a continuous variable) by subtracting it from its mean (see [2] for details on multiplicative terms).

Table 6.3 shows the four regression models of the gender and racial composition of the social issue discussion groups. Model 1 shows that, net of effects from other mediating variables, female students' groups have 0.314 more female discussion partners than male students' discussion groups. This result supports our first hypothesis asserting gender homophily that women tend to include other women in their network. To further examine how kinship exerts influences on gender homophily, we created the multiplicative term between female students and the proportion of kinship in the networks. Model 2 shows that the multiplicative term of female and kin portion is significant, indicating that the impacts of kinship on gender composition of discussion groups are significantly different between male and female students. In particular based on Model 2,

$$\textit{female proportion} = 0.327 + 0.316\,(\textit{female}) + 0.232\,(\textit{kin}) + (-0.342)(\textit{female} \times \textit{kin})$$

Model for male students is thus derived as

$$\textit{female proportion} = 0.327 + 0.316\,(0) + 0.232\,(\textit{kin}) + (-0.342)(0 \times \textit{kin})$$

$$\textit{female proportion} = 0.327 + 0.232(\textit{kin})$$

Model for female students is thus derived as

$$\begin{aligned}\textit{female proportion} &= 0.327 + 0.316\,(1) + 0.232\,(\textit{kin}) + (-0.342)(1 \times \textit{kin})\\ &= 0.327 + 0.232(\textit{kin})\end{aligned}$$

$$\textit{female proportion} = 0.643 + (-0.11)(\textit{kin})$$

Table 6.3 Unstandardized coefficients of OLS regression of gender and racial composition

Exogenous variables	Female prop. (I)	Female prop. (II)	Race IQV (I)	Race IQV (II)
Constant	.345*** (.085)	.327*** (.079)	−.104 (.107)	−.082 (.109)
Age (≥20)	.024 (.053)	.005 (.053)	−.144* (.062)	−.146* (.063)
Major_Agriculture (REF)	–	–	–	–
Major_Arts and Sciences	.50 (.067)	.062 (.066)	.062 (.076)	.065 (.075)
Major_Business	.053 (.074)	.064 (.072)	.044 (.083)	.033 (.082)
Major_Education	.034 (.070)	.013 (.069)	.003 (.080)	.006 (.080)
Major_Engineering	.070 (.093)	.105 (.093)	.162 (.108)	.148 (.107)
Major_others	.151 (.087)	.135 (.087)	−.007 (.100)	−.010 (.100)
Network size <3 (REF)	–	–	–	–
Network size = 3	−.002 (.082)	.016 (.081)	.176 (.110)	.157 (.115)
Network size = 4	.052 (.084)	.077 (.083)	.298* (.115)	.300* (.118)
Network size = 5	−.044 (.060)	−.024 (.060)	.268** (.094)	.253* (.010)
Gender_female	.314*** (.041)	.316*** (.040)	.020 (.047)	.011 (.047)
Race_White (REF)	–	–	–	–
Race_Black	−.020 (.073)	−.039 (.072)	.273** (.085)	.272** (.085)
Race_Hispanics	.015 (.065)	.001 (.064)	.407*** (.074)	.480*** (.079)
Race_others	−.009 (.065)	−.021 (.064)	.239** (.074)	.238** (.080)
Kin prop. (centered)	.027 (.063)	.232* (.098)	−.312*** (074)	−.340*** (.079)
Female × kin prop.	–	−.342** (.127)	–	–
Black × kin prop.	–	–	–	−.165 (.304)
Hispanics × kin prop.	–		–	.806* (.349)
Others × kin prop.	–	–	–	.035 (.373)
Model F ratio (df)/R^2	5.2***(14)/ 29.5%	5.51***(15)/ 32.3%	7.20***(14)/ 38.2%	6.38***(17)/ 40.4%

* $P < .05$; ** $P < .01$; *** $P < .001$ (two-tail test)

For male students, each unit of increase in kin proportion increases the number of females in their discussion network by 0.232 unit. On the contrary, for female students, each unit of increase in kinship reduces the number of females in their discussion network by 0.11 unit. In other words, while male students include females in their discussion network via kinship and family ties, female students recruit other females in their discussion network via other means, perhaps through social settings such as classmates, sorority parties, football/collegiate events etc. Combining results from both models, we found evidence supporting our second hypothesis: not only gender homophily in social issue discussion networks materializes (male students have more male partners, female students have more female discussion partners), but also gender difference in recruiting mechanism of discussion partners (male students rely on families/kinship ties to know females, whereas female students use nonfamily/kinship ties to recruit other females) exists.

Models 3 and 4 examine the racial composition of discussion networks, indicated by race IQV that ranges from 0 (complete racial homophily with only one race included) to 1 (complete racial diversity with all four groups having equal representation). Model 3 shows that network size, kin proportion, and students' race have significant impacts on racial diversity. Larger networks (those with 4 or 5 discussion partners) are more racially diverse than smaller ones (those with 1 or 2 discussion partners). Each unit of increase in kinship proportion reduces racial diversity by 0.342 units. Both results align with existing literature and conform to common senses: the larger the one's social networks, the more opportunities for inclusions of someone with different racial backgrounds. The more inclusion of someone from one's kinship (families and relatives), the more likely the persons are the same racial backgrounds with the egos. A bit surprising result is that age (dummy coded as those 20 years old or older) have lower racial diversity in their discission network compared with younger students. However, due to the sample that are young college students enrolling in an introduction course, we would not go too far on this finding.

Supporting our third hypothesis, Model 3 shows that race has significant impacts on the racial diversity of their discussion networks. Compared with white students, all minority students (blacks, Hispanics, and others) have more racially diverse networks. Considering that the racial makeup of the class (more than 80 percent students are white), which matches with the general students' racial makeup of the university, the higher racial homophily of white students compared with the minority students is not surprising. When it comes to finding and selecting conversation partners, white students simply have the number advantage that majority of others are whites. Conversely, minority students have a much restrictive pool to choose from if they are to select discussion partners within their respective minority groups. Consequentially, all minority groups have more partners from members of other racial groups than are white students, causing higher racial diversity in their groups.

Model 4 in Table 6.3 examines whether the effects of kinship on racial diversity of the discussion network vary significantly depending on racial identity of the students. It shows that the effect of kinship on racial diversity of discussion network for Hispanics is significantly different than it is for white students. In particular based on the model,

$$\begin{aligned}\text{Race IQV} = & -0.082 + 0.272\ (\text{black}) + 0.480\ (\text{Hispanics}) + 0.238\ (\text{others}) \\ & + (-0.340)(\text{kin}) + 0.806\ (\text{kin} \times \text{Hispanics})\end{aligned}$$

Model for white students is thus derived as

$$\text{Race IQV} = -0.082 + 0.272\ (0) + 0.480\ (0) + 0.238\ (0) + (-0.340)(\text{kin}) + 0.806\ \ (0)$$

$$\text{Race IQV} = -0.082 + (-0.340)(\text{kin})$$

Model for Hispanics students is thus derived as

$$\text{Race IQV} = -0.082 + 0.272\ (0) + 0.480\ (1) + 0.238\ (0) + (-0.340)(\text{kin}) + 0.806\ \ (\text{kin})$$

$$\text{Race IQV} = 0.398 + (0.466)(\text{kin})$$

Based on those results, we found that the extent to which kinship reduces one's social network racial diversity is a contingent one. For whites, black, and other minorities, kinship reduces racial diversity, which aligns with the literatures. On the contrary, for Hispanics, kinship increases racial diversity of their networks. Such result partially supports our fourth hypothesis that the effect of kinship on racial homophily is more pronounced for whites than it is for minorities. Here, we report such difference materializes between white and Hispanics. Perhaps this is due to the difference between Hispanics and non-Hispanics white, blacks, and other minorities in terms of their family racial makeups. Hispanic families may have a much higher racial diversity than non-Hispanics families, resulting in that adding kin/familial members to the discussion networks increases rather than decreases the racial diversity of their networks.

6.7 Conclusion

During this unprecedented time of covid-19 pandemic that wreaks havoc many social orders and social life, we investigate the social networking among college students using our innovative social issue generators that capture the most current social events. We use convenient sampling to collect data from 191 students who enrolled in a General Sociology course about their discussion partners on five critical social issues: Covid-19 and its related topics such as facemask, school closing, college sports, and other general issues such as race and inequality, and race and policing. On the average, each student names 4.43 conversation partners on those issues. We also look at the gender and racial composition of the social issue discussion networks. Overall, gender composition is slightly in favor of female, with almost 60 percent of the conversation partners being female. Racial homophily is present in our social issue discussion networks, the average IQV of racial composition is below 25 percent.

Gender homophily is obvious in our sample that female students have more females in their discussion network that do male students. To untangle the puzzles of gender homophily and kinship nomination in social networking, we created multiplicative terms in our regression models to examine whether men and women are different in terms of naming other family members in their network gender composition. Yes, such difference does exist. Including family members increases the number of women in men's discussion network, but it decreases the number of women in women's network. In other words, women include other women in their discussion network not through family ties, but through other social setting above and beyond family boundaries.

Regarding racial homophily, not only it is present in our sample, but also it is different between different racial groups. Racial homophily is significantly higher for white students than it is for minority students. We also look at how including family members affect such racial difference in racial homophily. The results show that while kinship nominations increase the racial homophily, one racial group: Hispanics do not follow. Increases in family members in their discussion networks actually decrease the racial homophily. We suspect such result may reflect the family structures of Hispanics, which is perhaps more multi-racial compared with other racial groups.

We contextualize our study in line of the debate of increasing social isolation and decreasing social connectivity for American's general public [31, 33]. Researchers use the important matter generator to report that American's social networking become smaller, and more condensed to include only family members from 1985 to 2004. Such finding inspires many studies and a great deal of debates to examine the issue of social isolation and social withdraws (reducing one's contacts to only family members). The historically unparalleled Covid-19 pandemic reignites the discussions of social isolation and reduced social connectivity. Do people stay connected and engaged with each other while the pandemic shuts down campuses, closing officers, reducing store hours and shopping activities, and sending everybody back home, locked up, and socially distanced from each other?

From what we observed, our student samples are very well connected, that they find plenty of conversation partners to discuss the pivotal social issues. Most likely, though, those conversations take place via ethernet supported by some social media platforms. However, we did not ask about the mode of communications in our social issue generator prob., leaving an important research topic to pursue. More cogently, do online computer support social networking provide the same social connectivity and engagement as those offline face to face interactions? One study [7] suggests that online social networking can deliver the same social connectivity as those offline interactions, provided that social competency and social presence are present in those online connectivities.

With our convenient sample of college students, we are not drawing direct comparison with the GSS core discussion network studies that are based on national probabilistic sample. Rather, we situate our discussion of the social issue discussion network in reference to the core discussion network studies to highlight a couple of contributions of our research: (1) we developed the innovative social issue generators to capture discussion networks around current and pivotal social issues. We believe such network prob., compared with the important matter prob. is more direct, more concurrent with the current and critical social issues, leaving little room for respondents' self-interpretations that can vary greatly. (2) our study of social networking is very timely as this unprecedented pandemic created widespread social isolation as the Covid-19 shuts down schools, stores, work organizations, and almost all gathering places. During the height of this pandemic, all but seven States in American implemented the most draconian measures of shelter in place, locking everybody back at home. Our study, though based on a convenient sample of college students, indicate that people strive to stay connected and engaged with each other even during such historically unparalleled time of isolation and despair. Such finding is very preliminary, requiring future studies with more representative sample to confirm.

We also found not only gender and racial homophily are very present in our social issue discussion networks, but also the relations between gender, race, homophily, and kinship nomination are much more complex than we originally thought. For men, increases of family members is associated with increases of women in in their discussion networks. For women, increases of family members reduces the number of women in in their discussion networks. The gender preference that women tend to name other women in their discussion network does not go through family member nominations, but instead rely on other mechanisms beyond the realm of families.

Our finding also alerts us to an important new research topic: kinship and racial homophily in one's social networks. In the past literatures, kinship increases racial homophily because one's family members and relatives are more of the same race, thus the more one's family members being included in one's social networks, the higher the racial homophily. Our finding shows that Hispanics have a pattern that is opposite of such assumption: the higher the kinship, the lower the racial homophily. What does it reveal? Our speculation is that Hispanic families are more multi-racial and multi-culture than are the others, consequentially including other family members increases, not decreases the racial diversity in their social network. Such assumption demands future studies with pertinent data, such as national probabilistic sample and ideally with longitudinal designs. Are Hispanic families more multi-racial than are other families? Are all the families move toward multi-racial and multi-cultural directions?

Due to Covid-19 and campus showdown, we surveyed our students regarding their social networking with our social issue generators, using the innovative online discourse platform ICAS. However, we did not ask our students the mode of their interactions with their network partners, whether it is through in person face to face to through online with various computer/smart phone supported social medias, leaving an important topic for future social network studies. For long, the ego-centric social networking presumes to take place with face to face, in person interactions. Even with the online social media social networking becomes increasingly common, those online interactions have not been main mode of social interactions, until the Covid-19 hit. Many spheres where those in personal ego-centric social networking occurs are becoming online and remote, shopping, meeting, classes, many work activities are moved from in person to online. Topics are due to examine the extent to which people rely on online interactions, with whom those online interactions tend to occur, on what topics, frequency, effectiveness, outcomes, and comparison between online and offline social networking.

Appendix: Basic Demographic Features of the Student Sample

Variables	Distribution	Dispersion
Gender	Men = 55 (28.8%) Women = 136 (71.2%)	IQV = 82.0%
Race	White = 154 (80.6%) Black = 11 (5.8%) Hispanics = 13 (6.8%) Others = 13 (6.8%)	IQV = 45.0%
College	Agriculture = 14 (7.3%) Arts and science = 75 (39.3%) Business = 37 (19.4%) Education = 42 (22%) Engineering = 10 (5.2%) Others = 13 (6.8%)	IQV = 89.6%
Age	$\overline{X} = 18.76$	$S = 2.29$
N	191	–

References

1. Acock AC, Hurlbert JS (1993) Social networks, marital status, and well-being. Soc Networks 15:309–334
2. Aiken LS, West SG (1991) Multiple regression: testing and interpreting interactions. Sage Publications, Inc., Newbury Park
3. Bailey S, Marsden PV (1999) Interpretation and interview context: examining the social survey name generator and using cognitive methods. Soc Networks 21(3):287–309
4. Barnes LL, Mendes de Leon CF, Wilson RS, Bienias JL, Evans DA (2004) Social resources and cognitive dcline in a population of older African Americans and Whites. Neurology 63: 2322–2326
5. Beggs JJ, Haines VA, Hurlbert JS (1996) Revisiting the rural-urban contrast: personal networks in nonmetropolitan and metropolitan settings. Rural Sociol 61:306–325
6. Bienenstock EJ, Bonacich P, Oliver M (1990) The effect of network density and homogeneity on attitude polarization. Soc Networks 12:153–172
7. Bouchillon B (2021) Social networking for social capital: the declining value of presence for trusting with age. Behav Inform Technol 41:1425–1438
8. Brewer DD (1995) The social structural basis of the organization of persons in memory. Hum Nat 6:379–403
9. Burt RS (1984) Network items and the general social survey. Soc Networks 6:293–339

10. Brummett BH, Barefoot JC, Siegler IC, Clapp-Channing NE, Lytle BL, Bosworth HB, Williams RB, Mark DB (2001) Characteristics of socially isolated patients with coronary artery disease who are at elevated risk for mortality. Psychosom Med 63:267–272
11. Campbell KE, Lee BA (1991) Name generators in surveys of personal networks. Soc Networks 13:203–221
12. Cohen S, Doyle W, Skoner DP, Rabin BS, Gwaltney JM (1997) Social ties and susceptibility to the common cold. JAMA J Am Med Assoc 277:1940–1944
13. Deng Z, Bonacich P (1991) Some effects of urbanism on Black networks. Soc Networks 13: 35–50
14. Fischer CS (2009) The 2004 GSS finding of shrunken social networks: an artifact? Am Sociol Rev 74(4):657–669. https://doi.org/10.1177/000312240907400408
15. Fischer CS (2011) Still connected: family and friends in America since 1970. Russel Sage Foundation, New York
16. Fischer CS, Shavit Y (1995) National differences in network density: Israel and the United States. Soc Networks 17:129–145
17. Fountain H (2006) The lonely American just got a bit lonelier. New York Times, July 2
18. Hawkley LC, Masi CM, Berry JD, Cacioppo JT (2006) Loneliness is a unique predictor of age-related differences in systolic blood pressure. Psychol Aging 21:152–164
19. Heikkinen R, Kauppinen M (2004) Depressive symptoms in late life: a 10-year follow-up. Arch Gerontol Geriatr 38:239–250
20. Huang G, Tausig M (1990) Network range in personal networks. Soc Networks 12:261–268
21. Hurlbert JS (1991) Social networks, social circles, and job satisfaction. Work Occup 4:415–430
22. Hurlbert JS, Acock AC (1990) The effects of marital status on the form and composition of social networks. Soc Sci Q 71:163–174
23. Hurlbert JS, Haines VA, Beggs JJ (2000) Core networks and tie activation: what kinds of routine networks allocate resources in nonroutine situations? Am Sociol Rev 65:598–618
24. Knoke D (1990) Networks of political action: toward theory construction. Soc Forces 68: 1041–1063
25. Knoke D, Yang S (2020) Social network analysis. SAGE Publications, Inc., Thousand Oaks
26. Kovacs B, Caplan N, Grob S, King M (2021) Social networks and loneliness during the COVID-19 pandemic. Socius. https://doi.org/10.1177/2378023120985254
27. Lazarsfeld PF, Merton RK (1954) Friendship as a social process: a substantive and methodological analysis. In: Berger M (ed) Freedom and control in modern society. Van Nostrand, New York, pp 18–66
28. Liao TF, Stevens G (1994) Spouses, homogamy, and social networks. Soc Forces 73:693–707
29. Lin N, Dumin M (1986) Access to occupations through social ties. Soc Networks 8(4):365–385
30. Liu Y, Lu D, Zhang Y, Li S, Liu X, Lin H (2010) The evolution of somatostatin in vertebrates. Gene 463(1–2):21–28
31. Marsden PV (1987) Core discussion networks of Americans. Am Sociol Rev 52(1):122–131
32. McPherson M, Smith-Lovin L, Cook JM (2001) Birds of a feather: Homophily in social networks. Annu Rev Sociol 27:4150444
33. McPherson M, Smith-Lovin L, Brashears ME (2006) Social isolation in America: changes in core discussion networks over two decades. Am Sociol Rev 71:353–375
34. McPherson M, Smith-Lovin L, Brashears ME (2009) Models and marginals: using survey evidence to study social networks. Am Sociol Rev 74(4):670–681
35. Moore G (1990) Structural determinants of men's and women's personal networks. Am Sociol Rev 55:726–735
36. Olds J, Schwartz RS (2009) The lonely American: drifting apart in the twenty-first century. Beacon Press, Boston
37. Paik A, Sanchagrin K (2013) Social isolation in America: an artifact. Am Sociol Rev 78(3): 339–360
38. Parigi P, Henson WJ (2014) Social isolation in America. Rev Soc 40:153–171

39. Pressman SD, Cohen S, Miller GE, Barkin A, Rabin BS, Treanor JJ (2005) Loneliness, social network size, and immune response to influenza vaccination in college freshmen. Health Psychol 24:297–306
40. Putnam RD (2001) Social capital: Measurement and consequences. Can J Policy Res 2:41–51
41. Ruan D (1998) The content of the general social survey discussion networks: an exploration of general social survey discussion name generator in a Chinese context. Soc Networks 20: 247–264
42. Seeman TE (2000) Health promoting effects of friends and family on health outcomes in older adults. Am J Health Promot 14:362–370
43. Sigman A (2009) Well connected? The biological implications of 'social networking'. Biologist 56(1):14–20
44. Simmel G, Levine DN (1971) On individuality and social forms: selected writings. University of Chicago Press, Chicago
45. Sirrianni J, Liu X, Adams D (2020) Agreement prediction of arguments in cyber argumentation for detecting stance polarity and intensity. ACL Anthology
46. Straits BC (1996) Ego-net diversity: same and cross-sex co-worker ties. Soc Networks 18:29–45
47. Thoits PA, Hewitt LN (2001) Volunteer work and well-being. J Health Soc Behav 42:115–131
48. Uchino BN, Cacioppo JT, Kiecolt-Glaser JK (1996) The relationship between social support and physiological processes: a review with emphasis on underlying mechanisms and implications for health. Psychol Bull 119:488–531
49. Van der Poel MG (1993) Delineating personal support networks. Soc Networks 15(1):49–70
50. Verbrugge LM (1977) The structure of adult friendship choices. Soc Forces 56:576–597
51. Wellman B, Wortley S (1990) Difference strokes from different folks: community ties and social support. Am J Sociol 96:558–588
52. Wilson RS, Krueger KR, Arnold SE, Schneider JA, Kelly JF, Barnes LL, Tang Y, Bennett DA (2007) Loneliness and risk of Alzheimer disease. Arch Gen Psychiatry 64:234–240
53. Yang S, Keller F, Zheng L (2016) Social network analysis. SAGE Publications, Inc. https://doi.org/10.4135/9781071802847

Index

S. Yang (ed.), *Social Network Analysis in Action*, Lecture Notes in Social Networks, https://doi.org/10.1007/978-3-031-66661-2

The manufacturer's authorised representative in the EU is Springer Nature Customer Service Centre GmbH, Europaplatz 3, 69115 Heidelberg, Germany. If you have any concerns regarding our products, please contact ProductSafety@springernature.com

Printed and bound by CPI Group (UK) Ltd, Croydon, CR0 4YY
01/10/2026
02225977-0004